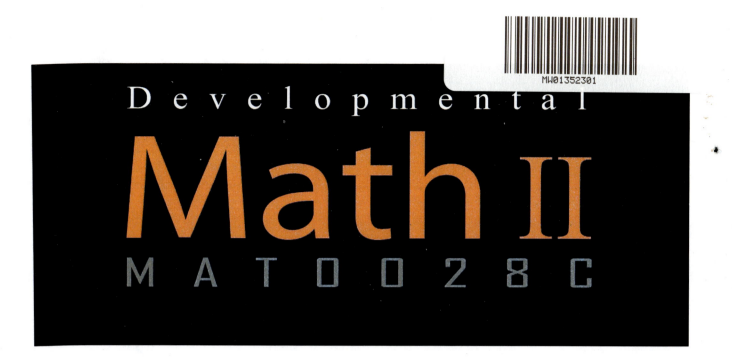

Second Edition

Al Groccia

VALENCIA COLLEGE

Published by Linus Learning

Ronkonkoma, NY 11779

Copyright © 2014 Linus Learning

All Rights Reserved.

ISBN 10: 1-60797-455-X

ISBN 13: 978-1-60797-455-0

No part of this publication may be reproduced, stored in a retrieval system, or transmitted, in any form or by any means, electronic, mechanical, photocopying, recording, or otherwise, without the prior permission of the publisher.

Printed in the United States of America.

This book is printed on acid-free paper.

Print Number   5   4   3   2   1

MAT0028C DEVELOPMENTAL MATH II STATE OF FLORIDA COMPETENCY LIST

COMPETENCY ID	MATHEMATICS CATEGORY	MATHEMATICS COMPETENCIES – UPPER	CONNECTION TO TEXTBOOK
MDECU1	Exponents & Polynomials	Applies the order of operations to evaluate algebraic expressions, including those with parentheses and exponents	Chapter 4
MDECU2	Exponents & Polynomials	Simplifies an expression with integer exponents	4.1
MDECU3	Exponents & Polynomials	Add, subtract, multiply, and divide polynomials. Division by monomials only. *(Does not include division by binomials)*	4.5, 4.6, 4.7
MDECU4	Factoring	Solve quadratic equations in one variable by factoring	5.6
MDECU5	Factoring	Factor polynomial expressions (GCF, grouping, trinomials, difference of squares)	5.1, 5.2, 5.3, 5.4, 5.5
MDECU6	Graphing	Graph linear equations using table of values, intercepts, slope intercept form	3.2, 3.3, 3.5
MDECU7	Linear Equations	Solve linear equations in one variable using manipulations guided by the rules of arithmetic and the properties of equality.	2.2
MDECU8	Linear Equations	Solve literal equations for a given variable with applications (geometry, motion [d = rt], simple interest [i = prt])	2.5
MDECU9	Radicals	Simplify radical expressions – square roots only	6.1, 6.2
MDECU10	Radicals	Adds, subtracts, and multiplies square roots of monomials	6.3, 6.4
MDECU11	Exponents & Polynomials	Convert between scientific notation and standard notation	4.3
MDECU12	Exponents & Polynomials	Solve application problems involving geometry (perimeter and area with algebraic expressions)	2.7
MDECU13	Graphing	Identifies the intercepts of a linear equation	3.3, 3.5
MDECU14	Graphing	Identify the slope of a line (from slope formula, graph, and equation)	3.4, 3.5
MDECU15	Linear Equations	Solve multi–step problems involving fractions and percentages (Include situations such as simple interest, tax, markups/markdowns, gratuities and commissions, fees, percent increase or decrease, percent error, expressing rent as a percentage of take–home pay)	2.7
MDECU16	Linear Equations	Solve linear inequalities in one variable and graph the solution set on a number line	2.6
MDECU17	Radicals	Rationalize the denominator (monomials only)	6.5
MDECU18	Radicals	Solve application problems involving geometry (Pythagorean Theorem)	6.1
MDECU19	Rationals	Recognize proportional relationships and solve problems involving rates and ratios	2.3
MDECU20	Rationals	Simplify, multiply, and divide rational expressions	5.7, 5.9
MDECU21	Rationals	Add and subtract rational expressions with monomial denominators	5.10
MDECU22	Rationals	Convert units of measurement across measurement systems	2.4

Table of Contents

CHAPTER 1:
Real Numbers 1

- 1.1 Displaying Information and Vocabulary 2
- 1.2 Fractions 7
- 1.3 Real Numbers 15
- 1.4 Adding/Subtracting Real Numbers 19
- 1.5 Multiplying/Dividing Real Numbers 25
- 1.6 Exponents and Order of Operations 28
- 1.7 Algebraic Expressions 33

CHAPTER 2:
Algebraic Expressions and Equations 39

- 2.1 Simplifying Algebraic Expressions 40
- 2.2 Solving Equations 44
- 2.3 Proportions 52
- 2.4 Convert Units of Measurement across Measurement Systems 58
- 2.5 Formulas 63
- 2.6 Solving Inequalities 68
- 2.7 Problem Solving 75

CHAPTER 3:

Graphing 91

3.1	Graphing Using the Rectangular Coordinate System.............	92
3.2	Graphing Linear Equations Using Table of Values................	98
3.3	Graphing Linear Equations Using Intercepts	105
3.4	Slope of a Line	109
3.5	Slope Intercept...	117

CHAPTER 4:

Polynomials 131

4.1	Properties of Exponents...	132
4.2	Zero and Negative Exponents	140
4.3	Scientific Notation..	146
4.4	Polynomials..	149
4.5	Adding/Subtracting Polynomials..	155
4.6	Multiplying Polynomials...	159
4.7	Division of Polynomials (Monomials)..................................	166

CHAPTER 5:

Factoring/Rational Expressions 175

5.1	Factoring by GCF and Grouping	176
5.2	Factoring Trinomials in the Form of $x^2 + bx + c$	185
5.3	Factoring Trinomials in the Form of $ax^2 + bx + c$	191
5.4	Factoring the Difference of Two Squares and Perfect Square Trinomials...	198

5.5	Factoring Using Multiple Methods	204
5.6	Solve Quadratic Equations by Factoring	207
5.7	Factoring Applications	212
5.8	Simplify Rational Expressions	219
5.9	Multiply and Divide Rational Expressions	224
5.10	Add and Subtract Rational Expressions with Monomial Denominators	231

CHAPTER 6:

Radicals 245

6.1	An Introduction to Square Roots	246
6.2	Simplifying Square Roots	253
6.3	Adding and Subtracting Radical Expressions	261
6.4	Multiplying and Dividing Radical Expressions	265
6.5	Rationalizing the Denominator (monomials only)	273
6.6	Solving Radical Equations	277
6.7	Higher Order Roots	283

Practice Problems 293

REAL NUMBERS

CHAPTER 1

- **1.1: DISPLAYING INFORMATION/VOCABULARY**
- **1.2: FRACTIONS**
- **1.3: REAL NUMBERS**
- **1.4: ADDING/ SUBTRACTING REAL NUMBERS**
- **1.5: MULTIPLYING/ DIVIDING REAL NUMBERS**
- **1.6: EXPONENTS AND ORDER OF OPERATIONS**
- **1.7: ALGEBRAIC EXPRESSIONS**

CHAPTER 1

1.1: DISPLAYING INFORMATION AND VOCABULARY

Why do we need Tables and Graphs?

Creating a Table

Table Title: <u>Movie Box Office Sales</u>

Source: www.leesmovieinfo.com

Movie Title	Box Office Sales (US and Canada) Rounded
Shrek	$268,000,000
The Notebook	$80,000,000
Star Wars	$461,000,000
Happy Gilmore	$39,000,000
The Godfather	$135,000,000
Rocky	$117,000,000

Creating a Bar Graph

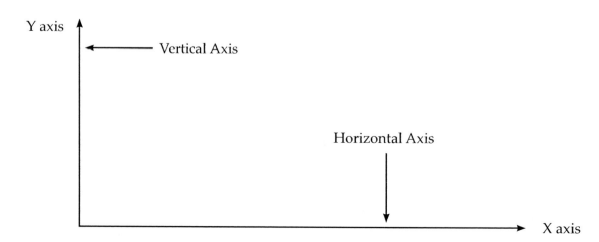

Creating a Line Graph

Sales for a Luxury Car Dealers in the United States during 2001	
Month	Sales
January	$2,500,000
June	$6,250,000
September	$750,000
December	$9,000,000

Why do we need Tables and Graphs?

_____
_____
_____
_____

CHAPTER 1

Vocabulary

Sum- is the result of addition.
Example: The sum of 5 and 7 is 12.

Difference- is the result of subtraction.
Example: The difference of 3 and 2 is 1.

Product- is the result of multiplication.
Example: The product of 4 and 7 is 28.

Quotient- is the result of division.
Example: The quotient of 12 and 3 is 4.

Notation

Multiplication Symbols

×   Times Sign (Due to possible confusing with the variable sign "x", this sign will not be used often.)

Example: $5 \times 4 = 20$

•   Raised dot

Example: $3 \cdot 2 = 6$

( )   Parentheses

Example: $(6)(7) = 42$

Division Symbols

÷   Division Sign

Example: $15 \div 3 = 5$

⟌   Long Division

Example: $4\overline{)24} = 6$

—   Fraction Bar

Example: $\frac{18}{2} = 9$

REAL NUMBERS 5

Variables, Expressions, and Equations

★ **Variables**- Letters that stand for numbers.
Example: If you don't know how much money you have, the unknown value can be represented by a variable like "x".

★ **Equation**- A mathematical sentence that contains the equality sign =.
Examples: $3 + 5 = 8$ or $x + 9 = 12$

★ **Algebraic Expression**- variables and/or numbers that can be combined with the operations of addition, subtraction, multiplication, and division.
Examples: $x + 7$ or $(xyz)/3$ or $10ac(5a)$
> Can be simplified or evaluated (not solved)

Constructing Tables

Example:

Movie Tickets cost $8 each.
How much would it cost to bring my friends to the movies?

We do not know how many friends are there.
We will use the variable "f" to represent friends.

We also do not know the total cost.
We will use the variable "c" to represent the total cost.

What will the formula be to represent this situation?
_____

Use this table to show the possible costs based on how many friends go to the movies?

f	c
1	8

1.1: INTRO LANGUAGE OF ALGEBRA PRACTICE PROBLEMS

1. Create a table and bar graph for the following information:

   The most played songs of 2006: (Source: www.nielsenmedia.com)

   "Be Without You" Mary J. Blige (395,995 times), "Unwritten" Natasha Bedingfield (336,276), "Temperature" Sean Paul (324,555), "Me & U" Cassie (312,073), "Hips Don't Lie" Shakira (308,903), "Promiscuous" Nelly Furtado (292,264).

   TABLE: BAR GRAPH:

2. Create a table and line graph for the following information:

   Here are the yearly enrollments for Valencia Community College:

   (Source: http://valenciacc.edu/IR/EnrollmentStatistics.cfm)

   1967: 567   1977: 24,483   1987: 54,515   1997: 48,503   2005: 53,806

   TABLE: LINE GRAPH:

3. Match the vocabulary word with the definition or symbols:

   ____ 1. Algebraic Expression   a. Result of addition
   ____ 2. Product   b. Letters that stand for numbers
   ____ 3. Division Symbols   c. Result of division
   ____ 4. Quotient   d. Algebraic equation for the cost of 3 unknown priced tickets with a coupon for $2 off is $10.
   ____ 5. $3(x) - 2 = 10$ e. Result of subtraction
   ____ 6. Difference   f. Variables and/or numbers combined with arithmetic operations
   ____ 7. Multiplication Symbols   g. $\times, \cdot, (\ )$
   ____ 8. Variables   h. $\div, \overline{)}, -$
   ____ 9. Sum   i. Result of multiplication
   ____ 10. Equation   j. Mathematical Sentence with the equality sign =

REAL NUMBERS

1.2:
FRACTIONS

What is a FRACTION and why do we need them?

Factorization- to express a number as the product of two or more numbers.

Example: 24 = 2·12   or   24 = 3·8   or   24 = 4·6   or   24 = 2·2·2·3

Prime number- a whole number greater than 1 that has only itself and 1 as factors.

The first ten prime numbers are 2, 3, 5, 7, 11, 13, 17, 19, 23, and 29.

Composite number- a whole number greater than 1 that is not prime.

The first ten composite numbers are 4, 6, 8, 9, 10, 12, 14, 15, 16 and 18

Prime Factorization- every composite number can be factored into the product of two or more prime numbers.

Example: Find the prime factorization of 210

		210	
	21		10
3	7	2	5

The prime factors are 2, 3, 5, 7.

Practice Examples:

1. Find the prime factorization of 256:

2. Find the prime factorization of 189:

Meaning of Fractions

Example of a fraction: $\frac{1}{2}$

1	Numerator
/	Fraction bar
2	Denominator

Special Fraction Forms: For any nonzero number a,

$\frac{a}{a} = 1$ 	ex. $\frac{5}{5} = 1$ 	$\frac{1000}{1000} = 1$

$\frac{a}{1} = a$ 	ex. $\frac{7}{1} = 1$ 	$\frac{256}{1} = 256$

$\frac{0}{a} = 0$ 	ex. $\frac{0}{9} = 0$ 	$\frac{0}{34} = 0$

$\frac{a}{0} = undefined$ 	ex. $\frac{3}{0} = undefined$ 	$\frac{45}{0} = undefined$

Simplifying Fractions - a fraction in its simplest form, or lowest terms, is when the numerator and denominator have no common factors other than 1 and themselves.

Example:

Simplify $\frac{30}{36}$

Solution:

Find the prime factorizations of the numerator and denominator. If the numerator and denominator have common factors, simplify.

$$\frac{30}{36} = \frac{2 \cdot 3 \cdot 5}{2 \cdot 2 \cdot 3 \cdot 3} = \frac{\cancel{2} \cdot \cancel{3} \cdot 5}{\cancel{2} \cdot 2 \cdot \cancel{3} \cdot 3} = \frac{5}{2 \cdot 3} = \frac{5}{6}$$

Examples:

Simplify:

1. $\frac{12}{18}$	2. $\frac{33}{40}$	3. $\frac{24}{56}$	4. $\frac{63}{42}$

REAL NUMBERS

Mixed Numbers- represent the sum of a whole number and a fraction.

Ex. $5\frac{3}{4}$

Improper Fraction- a fraction a which the numerator is greater than or equal to denominator.

Ex. $\frac{23}{4}$

Converting a mixed fraction to an improper fraction:

Example:

$5\frac{3}{4}$

Solution:

Multiply the denominator by the whole number, and then add the result to the numerator. That number becomes the numerator with the original denominator.

$5\frac{3}{4}$, $4 \times 5 = 20$, $20 + 3 = 23$, $\frac{23}{4}$

Examples:

Convert mixed numbers to improper fractions:

1. $1\frac{1}{2}$
2. $3\frac{2}{3}$
3. $7\frac{5}{6}$

Converting an improper fraction to a mixed number:

Example:

$\frac{23}{4}$

Solution:

Divide the denominator by the numerator. The result will be the whole number, the remainder will be the numerator, and the original denominator will be the denominator.

$\frac{23}{4} = 4\overline{)23}^{\,5}$ remainder 3 $5\frac{3}{4}$

Examples:

Convert the improper fractions to a mixed numbers:

1. $\frac{7}{2}$
2. $\frac{42}{5}$
3. $\frac{101}{12}$

CHAPTER 1

Multiplying Fractions- multiplying the numerators and the denominators.

Example:

Multiply $\frac{7}{8} \cdot \frac{3}{5}$. *Remember to simplify your answer if needed.*

SOLUTION:

$$\frac{7}{8} \cdot \frac{3}{5} = \frac{7 \cdot 3}{8 \cdot 5} = \frac{21}{40}$$

Examples:

Multiply:

1. $\frac{2}{3} \cdot \frac{4}{5} =$

2. $\frac{2}{5} \cdot \frac{3}{4} =$

3. $\frac{15}{8} \cdot \frac{4}{3} =$

4. $1\frac{1}{2} \cdot 2\frac{2}{3} =$

Dividing Fractions- multiply the first fraction by the reciprocal of the second. Remember to simplify your answer.

Example:

Divide: $\frac{1}{7} \div \frac{2}{5}$. *Remember to simplify your answer if needed.*

SOLUTION:

$$\frac{1}{7} \div \frac{2}{5} = \frac{1}{7} \cdot \frac{5}{2} = \frac{1 \cdot 5}{7 \cdot 2} = \frac{5}{14}$$

Examples:

Divide:

1. $\frac{2}{3} \div \frac{4}{5} =$

2. $\frac{2}{5} \div \frac{3}{4} =$

3. $\frac{8}{15} \div \frac{2}{3} =$

4. $2\frac{3}{4} \div 6\frac{3}{5} =$

REAL NUMBERS

Adding and Subtracting Fractions- in order to add or subtract fractions, they must have the same denominator.

To add (or subtract) two fractions with same denominator, add (or subtract) their numerators and write the sum (or difference) over the common denominator.

Examples:

1. $\dfrac{2}{7} + \dfrac{3}{7} =$

2. $\dfrac{10}{13} - \dfrac{4}{13} =$

Solutions:

1. $\dfrac{2}{7} + \dfrac{3}{7} = \dfrac{2+3}{7} = \dfrac{5}{7}$

2. $\dfrac{10}{13} - \dfrac{4}{13} = \dfrac{10-4}{13} = \dfrac{6}{13}$

To add (or subtract) two fractions with unlike denominator, find the least common denominator, convert both fractions to the common denominator, then add (or subtract) their numerators and write the sum (or difference) over the common denominator.

Least Common Denominator (LCD)- for a set of fractions the LCD is the smallest number each denominator will exactly divide (divide with no remainder).

To find the LCD, find the prime factorization of both denominators and use each prime factor the greatest number of times it appears in any one factorization.

Example:

Find the LCD of 10 and 28

Solution:

First find the prime factorization of each number:

$10 = 2 \cdot 5$

$28 = 2 \cdot 2 \cdot 7$

Then use each prime factor the greatest amount of times it appears in each number:

$$2 \cdot 2 \cdot 5 \cdot 7 = 140$$

Examples:

1. Find the LCD of 15 and 20

2. Find the LCD of 8 and 12

CHAPTER 1

Adding and Subtracting Fractions with unlike denominators

Example:

1. $\dfrac{2}{9} + \dfrac{7}{21} =$

2. $3\dfrac{5}{8} - 1\dfrac{1}{20} =$

Solutions:

The LCD is: $\begin{aligned} 9 &= 3 \cdot 3 \\ 21 &= 3 \cdot 7 \end{aligned}$ $\quad 3 \cdot 3 \cdot 7 = 63$

$$\dfrac{2}{9} \cdot \dfrac{7}{7} = \dfrac{14}{63}$$
$$+\dfrac{7}{21} \cdot \dfrac{3}{3} = \dfrac{21}{63}$$
$$\dfrac{35}{63} = \dfrac{5}{7}$$

The LCD is: $\begin{aligned} 8 &= 2 \cdot 2 \cdot 2 \\ 20 &= 2 \cdot 2 \cdot 5 \end{aligned}$ $\quad 2 \cdot 2 \cdot 2 \cdot 5 = 40$

$$3\dfrac{5}{8} \cdot \dfrac{5}{5} = 3\dfrac{25}{40}$$
$$-1\dfrac{1}{20} \cdot \dfrac{2}{2} = 1\dfrac{2}{40}$$
$$2\dfrac{23}{40}$$

Examples:

Solve:

1. $\dfrac{5}{7} + \dfrac{3}{7} =$

2. $\dfrac{6}{15} - \dfrac{2}{9} =$

3. $5\dfrac{3}{10} + 2\dfrac{5}{14} =$

4. $8\dfrac{2}{9} - 7\dfrac{2}{3} =$

What is a FRACTION and why do we need them?

_____

_____

1.2: FRACTIONS PRACTICE PROBLEMS

1. Find the prime factorization of 184:

2. Find the prime factorization of 212:

3. Simplify: $\dfrac{1976}{1976}$ _____

4. Simplify: $\dfrac{0}{7}$ _____

5. Simplify: $\dfrac{14}{0}$ _____

6. Simplify: $\dfrac{20}{36}$ _____

7. Simplify: $\dfrac{120}{300}$ _____

8. Simplify: $\dfrac{21}{70}$ _____

Convert the mixed number to an improper fraction:

9. $1\dfrac{2}{3}$ _____

10. $5\dfrac{4}{7}$ _____

11. $3\dfrac{1}{5}$ _____

Convert the improper fraction to a mixed number:

12. $\dfrac{8}{3}$ _____

13. $\dfrac{80}{7}$ _____

14. $\dfrac{123}{6}$ _____

Multiply:

15. $\dfrac{2}{5} \cdot \dfrac{3}{7} =$ _____

16. $\dfrac{25}{9} \cdot \dfrac{3}{5} =$ _____

17. $2\dfrac{1}{2} \cdot 3\dfrac{3}{5} =$ _____

Divide:

18. $\dfrac{3}{7} \div \dfrac{5}{6} =$ _____

19. $\dfrac{9}{40} \div \dfrac{5}{8} =$ _____

20. $5\dfrac{5}{7} \div 2\dfrac{6}{7} =$ _____

21. Find the LCD of 18 and 24

22. Add: $\dfrac{2}{7} + \dfrac{4}{7} =$ _____

23. Subtract: $\dfrac{9}{15} - \dfrac{4}{15} =$ _____

Solve:

24. $\dfrac{7}{12} - \dfrac{3}{8} =$ _____

25. $2\dfrac{4}{15} + 3\dfrac{6}{25} =$ _____

26. $10\dfrac{1}{6} - 8\dfrac{11}{20} =$ _____

1.3: REAL NUMBERS

What are REAL NUMBERS?

<u>Set</u>- collection of numbers, the symbol used is: { }

Natural Numbers
{1, 2, 3, 4...}   *Read as "the set containing 1, 2, 3, 4 and so on."*

Whole Numbers
{0, 1, 2, 3, 4...}   *Read as "the set containing 0, 1, 2, 3, 4 and so on."*

Integers
{... –3, –2, –1, 0, 1, 2, 3...}   Read as "the set containing -3, -2, -1, 0, 1, 2, 3 and so on in both directions."

Rational Numbers
Any number that can be written as a fraction with integer numerator and nonzero integer denominator.

Examples: $\frac{1}{2}$, $-3\frac{1}{4}$, $\frac{5}{3}$, 0.25, 0.333...., -7

<u>Irrational Numbers</u>- nonterminating, nonrepeating decimal.

Examples: 1.25987495..., π, $\sqrt{2}$

<u>Real Numbers</u>- rational and irrational numbers, represented as points on the number line.

Real Number	
Rational Number ↓	Irrational Number
Integer ↓	
Whole Number ↓	
Natural Number	

CHAPTER 1

Classify the following numbers:

7	
-5	
$\frac{5}{3}$	
$-\pi$	

Real Number Line

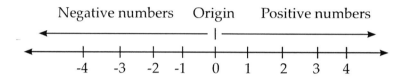

Graph the following on the number line:

$-2,\ 0,\ \frac{5}{2},\ -3\frac{1}{4},\ -1.5$

Inequality Signs:
< "Less Than", > "Greater Than"
Fill in the blanks with < or >:

-4		4
-2		-3
-5		-4
1.09		1.1
$-\frac{5}{2}$		$-\frac{3}{2}$

<u>**Opposite-**</u> two numbers that are at the same distance from 0.
Example: 4 and −4   −3 and 3
<u>**Absolute Value:**</u> the distance from 0. Symbol: | |

Example: $|5|=5,\ |-3|=3,\ -|-7|=-7$

Fill in the blanks with < or > or =

$|-4|$		4
$-(-5)$		-3
$-|-10|$		10

$-(9)$		8
$\left\|-\dfrac{1}{2}\right\|$		$-\left(-\dfrac{1}{3}\right)$

What are REAL NUMBERS?

_____
_____
_____

Set operations including union and intersection of sets

<u>Element of a set</u> - number, letter, or object in a set.
Example: If A = {1, 2, 3, 4}, the elements of set A are 1, 2, 3, 4

<u>Union of sets</u> (symbol: $\cup$) - combining all of the elements of two or more sets
Example: If A = {1,2,3,4} and B = {2,4,6}, the union of sets A and B is
$\quad A \cup B = \{1,2,3,4,6\}$

<u>Intersection of sets</u> (symbol: $\cap$) - the common elements of two or more sets
Example: If A = {1,2,3,4} and B = {2,4,6}, the intersection of sets A and B is
$\quad A \cap B = \{2,4\}$

<u>Empty set</u> - set containing no elements
$\quad$ (example: { })

Examples:

1. Let A = {2,4,6,8,10} and B = {4,8,16}.
   Find $A \cup B$:
   Find $A \cap B$:

2. Let C = {1,3,5,7,9,10} and D = {1,2,3,4,5,6}.
   Find $C \cup D$:
   Find $C \cap D$:

3. Let E = {1,2,3} and F = {4,5,6}.
   Find $E \cup F$:
   Find $E \cap F$:

1.3: REAL NUMBERS PRACTICE PROBLEMS

1. Classify the following numbers as: Natural, Whole, Integral, Rational, Irrational. Real numbers may fall into more than 1 category.

   a. $-\pi$ _____

   b. -10 _____

   c. $-\dfrac{9}{2}$ _____

   d. $23.458976975....$ _____

   e. 0 _____

   f. $\sqrt{9}$ _____

2. Create a number line and graph the following on the number line:

   $4,\ 0,\ -1,\ -5\dfrac{1}{2},\ \dfrac{9}{4},\ -2.75,\ \pi$

3. Complete the table with: < Less Than or > Greater Than or = Equal to

a.	-17		-18
b.	3.001		3.01
c.	$-\dfrac{8}{3}$		$-\dfrac{9}{3}$
d.	$|-6|$		$|6|$
e.	$-(-2)$		$-(2)$
f.	$-|-23|$		$-|23|$
g.	$-|-5|$		$-|5|$
h.	$\left|\dfrac{4}{5}\right|$		$-\left(-\dfrac{7}{10}\right)$
i.	$-(0.003)$		$-|0.004|$

4. Let A = {1, 5, 10, 20} and B = {5, 10, 15, 20}

   Find $A \cup B$:   Find $A \cap B$:

1.4: ADDING/SUBTRACTING REAL NUMBERS

How do we ADD REAL NUMBERS?

Adding Real Numbers

Signed Numbers- positive and negative numbers

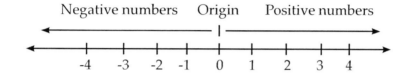

Adding Two Numbers with the Same Sign:

1. Two positive numbers: add them and keep the sign.
   Example: 10 + 3 = 13   15 + 12 = 27

2. Two negative numbers: add them and keep the sign.
   Example: -2 + (-3) = -5   -1 + (-7) = -8

Examples:

1.   -22 + (-13) =

2.   -1.23 + (-2.45) =

3. $-\dfrac{1}{4} + \left(-\dfrac{1}{2}\right) =$

CHAPTER 1

Adding Two Numbers with Different Signs

Subtract the numbers and take the sign of the larger number (without the sign).
Example: −5 + 2 = -3   4 + (-1) = 3

Examples:

1. −30 + 10 =

2. 5.4 + (-2.2) =

3. $-\dfrac{7}{30} + \left(\dfrac{1}{5}\right) =$

4. −20 + 5 + (−12) + (−3) + 7 =

5. (−7 + 8) + 2 + (−13 + 12) =

Properties of Addition

Commutative Property of Addition - Numbers can be added in any order and the sum is always the same.

Example: $2 + 3 = 3 + 2$

Associative Property of Addition - when we add more than two numbers the grouping of the numbers does not change the sum.

Example: $(2 + 3) + 4 = 2 + (3 + 4)$

Addition Property of 0 - when 0 is added to any number, the number does not change.

Example: $5 + 0 = 5$

Addition Property of Opposites - the sum of a number and its opposite is 0.

Example: $7 + (-7) = 0$

How do we ADD REAL NUMBERS?

_____

_____

_____

How do we SUBTRACT REAL NUMBERS?
Subtracting Real Numbers

The opposite of a negative number is a positive number.

Example: −(−5) is 5

Examples:

1. −(−7) =
2. −(−g) =
3. −|−15| =

Subtraction of Real Numbers

Change the subtraction sign to an addition sign and take the opposite of the following number. Then follow the addition rules.

Example:

−5 − 3 =

SOLUTION:

Change the subtraction sign to addition and take the opposite of the following number.

−5 − 3 =

−5 + (−3) =

Then follow the addition rules.

−5 + (−3) = −8

Example:

7 − (-9) =

SOLUTION:

Change the subtraction sign to addition and take the opposite of the following number.

7 − (-9) =

7 + 9 =

Follow the addition rules.

7 + 9 = 16

Examples:

1. $-10 - 7 =$

2. $50 - 85 =$

3. $\dfrac{1}{3} - \left(-\dfrac{2}{9}\right) =$

4. $-7 - 4 + 10 - (-5) =$

5. **Water level.** In one week, the water level in a storage tank went from 25 feet above normal to 12 feet below. Find the change in the water level.

How do we SUBTRACT REAL NUMBERS?

_____

_____

1.4: ADDING/SUBTRACTING REAL NUMBERS PRACTICE PROBLEMS

Solve the following problems:

1. a. 10 + (-2) = _____ b. -4 + (-3) = _____ c. -12 + 6 = _____

2. a. -3.4 + (-2.15) = _____ b. $-\dfrac{2}{3}+\left(\dfrac{3}{5}\right)=$ _____ c. $5\dfrac{3}{4}+\left(-1\dfrac{1}{3}\right)=$ _____

3. a. -5 – 5 = _____ b. 10 - (-4) = _____ c. -3 – (-8) = _____

4. a. 6.25 – 10.75 = _____ b. $-\dfrac{2}{7}-\left(\dfrac{1}{2}\right)=$ _____ c. $-4\dfrac{5}{6}-\left(-2\dfrac{3}{4}\right)=$ _____

5. a. 7 + (-2) + (-3) + 8 + (-3) = _____ b. -2.3 + (-4.2) + 1.24 + (-0.4) = _____

6. a. 15 - (-3) + (-6) - 21 - (-1) = _____ b. $-\dfrac{1}{3}-\left(\dfrac{3}{4}\right)-\left(-\dfrac{5}{6}\right)=$ _____

7. Sam is on the fifth floor of a building and needs to use the restroom located on the third floor of the basement (3 floors below ground level). How many floors must he travel to arrive at the restroom? Give the equation and the solution of this problem.

1.5: MULTIPLYING AND DIVIDING REAL NUMBERS

How do we MULTIPLY/DIVIDE REAL NUMBERS?
Multiplying or Dividing Real Numbers

If the signs are the same, the solution is positive.

Examples: $2 \cdot 3 = 6$ $\qquad \dfrac{10}{5} = 2 \qquad -\dfrac{2}{3} \cdot -\dfrac{4}{5} = \dfrac{8}{15} \qquad \dfrac{-3.3}{-3} = 1.1$

If the signs are different, the solution is negative.

Examples: $-5 \cdot 7 = -35 \qquad \dfrac{-20}{4} = -5 \qquad \dfrac{1}{4} \cdot -\dfrac{3}{7} = -\dfrac{3}{28} \qquad \dfrac{2.8}{-2} = -1.4$

Examples:

1. $7(-12) =$

2. $\dfrac{-100}{-5} =$

3. $-\dfrac{5}{7} \cdot -\dfrac{3}{10} =$

4. $\dfrac{2.25}{-5} =$

5. $-\dfrac{1}{2} \div \dfrac{3}{10} =$

6. $(-3)(-2)(5)(-4) =$

Properties of Multiplication

<u>Commutative Property of Multiplication</u>- changing the order of multiplication does not affect the product.
Example: 2(3) = 3(2)

<u>Associative Property of Multiplication</u> - changing the grouping in multiplication does not affect the result.
Example: (5 · 3)2 = 5(3 · 2)

<u>Multiplication Property of 0</u>- the product of 0 and any number is 0.
Example: 4(0) = 0

<u>Multiplication Property of 1</u>- Multiplying a number by 1 does not change the number.
Example: 9(1) = 9

<u>Multiplicative Inverses</u>- the product of any number and its multiplicative inverse (reciprocal) is 1.
Example: $8\left(\frac{1}{8}\right) = 1$

Properties of Division

<u>Dividing by 1</u>- Dividing a number by 1 does not change the number.
Example: $\frac{7}{1} = 7$

<u>Dividing by itself</u>- the quotient of dividing a number by itself is 1.
Example: $\frac{6}{6} = 1$

<u>Division with 0</u>- the quotient of dividing by 0 is undefined. Dividing 0 by any number (except 0) is 0.
Example: $\frac{4}{0} = undefined$ $\frac{0}{4} = 0$

How do we MULTIPLY/DIVIDE REAL NUMBERS?

1.5: MULTIPLYING/DIVIDING REAL NUMBERS PRACTICE PROBLEMS

Solve the following problems

1. a. $4(-3) =$ _____ b. $-15(-4) =$ _____ c. $-\dfrac{5}{6}\left(\dfrac{3}{10}\right) =$ _____

2. a. $\dfrac{-30}{-3} =$ _____ b. $\dfrac{-72}{4} =$ _____ c. $\dfrac{2.25}{-0.5} =$ _____

3. a. $1.2(-3.5)$ _____ b. $-\dfrac{7}{12} \cdot \dfrac{2}{5} =$ _____ c. $\dfrac{-\dfrac{2}{3}}{-\dfrac{5}{12}} =$ _____

4. a. $(-1)(-6)(-3)(-4) =$ _____ b. $\left(-\dfrac{1}{2}\right)\left(\dfrac{6}{7}\right)\left(-\dfrac{3}{8}\right)(-2) =$ _____

5. Create a word problem in which to find the solution you need to multiply or divide negative and positive numbers.

_____
_____
_____
_____
_____

1.6: EXPONENTS AND ORDER OF OPERATIONS

Exponents- used to indicate repeated multiplication

2^3 2 is the base and 3 is the exponent.

$$2^3 = 2 \cdot 2 \cdot 2 = 8$$

Examples:

Write each expression using exponents:

1. $5 \cdot 5 \cdot 5 \cdot 5 =$

2. $(-3)(-3)(-3)(-3)(-3)(-3) =$ _____

3. fourteen cubed =

4. $7 \cdot 7 \cdot 7 \cdot 12 \cdot 12 =$ (Figure out a pattern rather than calculating.) _____

5. $b \cdot b \cdot b \cdot b \cdot b \cdot b \cdot b \cdot b =$

6. $\dfrac{4}{3} \cdot \pi \cdot r \cdot r \cdot r =$ _____

Expand each expression and find the value:

1. $3^4 =$

2. $(-2)^3 =$ _____

3. $7^1 =$

4. $(-1)^{100} =$ _____

5. $\left(-\dfrac{1}{2}\right)^3 =$

6. $(0.4)^2 =$ _____

Evaluate or simplify:

1. -3^2 2. $(-3)^2$ 3. $-(3)^2$

Explain the differences _____

What are the steps for ORDER OF OPERATIONS?
Order of Operations

1. Grouping symbols (Parentheses)

    Grouping Symbols: $\{\,[\,(\,)\,]\,\}$

    Absolute Value: $|\ |$

    Square root: $\sqrt{}$

    <u>Note</u>: Fraction bar $\dfrac{(\ )}{(\ )}$ (First simplify numerator and denominator, then divide.)

2. **Exponents**

3. **Multiplication/ Division (Left to right)**

4. **Addition/ Subtraction (Left to right)**

Examples:

1. $2 \cdot 3^2 - 5 =$

2. $160 \div (-4) - 6(-2)^3 =$

CHAPTER 1

3. $100 \div 2 \cdot 5 - 12 + 4 =$

4. $-4\left[-2 - 3\left(4 - 8^2\right)\right] - 2 =$

5. $\dfrac{-3(3+2)+5}{8-3(-4)} =$

6. $10|2-5|-2^5 =$

7. $\dfrac{-|4-7|}{|4-7|} =$

8. $-4\left|3^2 - 5\right| + \left|-4 + 7(2)\right| - |-3| =$

What are the steps for ORDER OF OPERATIONS?

_____
_____
_____
_____

The Mean (Average)

Arithmetic mean (average)- divide the sum of the values by the number of values.

Example:

What is your test average if your test scores were?

Test 1: 90

Test 2: 80

Test 3: 100

Test 4: 70

32 CHAPTER 1

1.6: EXPONENTS/ORDER OF OPERATIONS PRACTICE PROBLEMS

Write each expression using exponents:

1. a. $2 \cdot 2 \cdot 2 \cdot 2 \cdot 2 \cdot 2 \cdot 2 =$ _____   b. $(-1)(-1)(-1)(4)(4)(-7) =$ _____

2. a. $-2 \cdot 5 \cdot -2 \cdot 5 \cdot 5 \cdot x \cdot y \cdot y =$ _____   b. $3 \cdot 3 \cdot 3 \cdot a \cdot a \cdot a \cdot a \cdot b \cdot b \cdot c =$ _____

Expand each expression and find the value:

3. a. $4^3 =$ _____   b. $(-3)^5 =$ _____

4. a. $0^1 =$ _____   b. $(-1)^{31} =$ (Don't expand, just solve) _____

5. a. $(1.2)^3 =$ _____   b. $\left(-\dfrac{2}{3}\right)^4 =$ _____

6. a. $(-5)^2 =$ _____   b. $-5^2 =$ _____

Solve:

7. a. $5(-2)^3 - 3(-4) - 6^2 =$ _____   b. $60 \div (3)(-2) - 4 + 10 =$ _____

8. a. $80 \div (2 \cdot 4) - (5^2 - 9) =$ _____   b. $-[8 - 5(3 - 2^2)] - 7 =$ _____

9. a. $\dfrac{2(-1-4)^2 - 6}{8 - |2 - 6|} =$ _____   b. $-|-3||7 - 5| - 4^2 =$ _____

10. What is the basketball point average if the points per game were?: _____
    Game 1: 14   Game 2: 18   Game 3: 20   Game 4: 28   Game 5: 21   Game 6: 32

1.7:
ALGEBRAIC EXPRESSIONS

What are Algebraic EXPRESSIONS and why do we use them?

Algebraic expression- variables and/or numbers combined with arithmetical operations.
Example: $3x^2 y + 5x - 7$

Term- part of **algebraic** expression that is separated by addition or subtraction.
Example: The terms of the algebraic expression $3x^2 y + 5x - 7$ are: $3x^2 y$, $5x$, and -7.

Coefficient- a numerical quantity or constant placed before and multiplying the variable in an algebraic term.
Example: The coefficient of the terms $3x^2 y$, $5x$, and -7 are: 3, 5, -7.

Polynomial- an expression consisting of the sum of two or more terms each of which is the product of a constant and variable raised to an integral power.

Polynomial equation- equation that contains polynomials.
Example: $5x^2 - 7 = 2x + 5$

Example:

Given the polynomial expression: $2x^2 - 6x + 9$
What are the terms? _____

What are the coefficients? _____

Example:

Given the polynomial expression: $x^3 - 4x^2 y + 5xy - 6$
What are the terms? _____

What are the coefficients? _____

Translating Words to Symbols

Addition	
Sum of x and 5	x + 5
f plus 9	f + 9
5 added to b	b + 5
8 more than r	r + 8
Y increased by h	Y + h
Exceed 6 by u	6 + u

Subtraction	
Difference of 5 and h	5 - h
100 minus b	100 − b
25 less than w	w − 25
8 decreased by J	8 − J
B reduced by 7	B − 7
10 subtracted from v	v - 10
L less 4	L - 4

Multiplication	
product of 5 and x	5x
25 times g	25g
twice w	2w
triple x	3x
$\frac{1}{2}$ of P	$\frac{1}{2}$P
x squared	$x \cdot x = x^2$

Division	
quotient of x and 10	$\frac{x}{10}$
W divided by S	$\frac{w}{s}$
ratio of 5 to b	$\frac{5}{b}$
D split into 6 parts	$\frac{D}{6}$

Equals	
X plus 5 equals 7	X + 5 = 7
Twice b results in 10	2b = 10
6 times a number is 12	6x = 12

Special	
two consecutive numbers	x and (x +1)
two consecutive even (or odd) numbers	x and (x + 2)

Examples:

Translate into an algebraic expression or algebraic equation

1. Five times b plus twice w:

2. Twelve less than the product of 4 and Y:

3. If 4 times a number is increased by 13, the result is 40 less than the square of the number:

4. The sum of a number and 9 is 5 more than twice the number:

5. Seven times the sum of a number and 3 is equal to 15:

6. If 10 times a number is decreased by 25, the result is 12 less than twice the number:

7. The product of a number and the next consecutive even number is 6:

Evaluating Algebraic Expressions

To evaluate an algebraic expression, substitute given numbers for each variable and do the necessary calculations (order of operations/simplifying).

Example:

Evaluate the expression $2x - 4y$ given that $x = 2$ and $y = -1$.

SOLUTION:

$2x - 4y$

Substitute the values for x and y into the expression

$2(2) - 4(-1)$

Evaluate the problem

$2(2) - 4(-1)$

$4 + 4$

8

Examples:

1. Evaluate the expression $-w^2 + 6w - 5$ when $w = -3$:

2. Evaluate the expression $9xy - z^2$ when $x = -3$, $y = 4$, $z = -6$:

3. Evaluate the expression $\dfrac{x-3y}{y^3}$ when x = -4, y = -1:

4. Evaluate the expression $|x-2y|+3|x|$ when x = -2, y = 3:

What are ALBEGRAIC EXPRESSIONS and why do we use them?

_____
_____
_____
_____

1.7: ALGEBRAIC EXPRESSIONS PRACTICE PROBLEMS

1. Given the algebraic expression: $2x^5 - x^3y^7 + 4x^5y + y - 8$:

   a. What are the terms? _____

   b. What are the coefficients? _____

Translate the following phrases into algebraic expressions or algebraic equations

2. The sum of a number and 5: _____

3. The product of a number and 3 less than 7: _____

4. If 7 times a number is decreased by 2, the result is 10 less than the twice the number:

   _____

5. Four times the sum of a number and 8 is equal to 40:

   _____

6. The sum of a number and the square of a number is equal to 5 less than twice the number:

   _____

7. The product of a number and the next consecutive odd number is 10:

   _____

8. Evaluate the expression $\dfrac{2a}{b} - c^2$ when: a = 6, b = -4, c = -3: _____

9. Evaluate the expression $x^3 - x^2 - x$ when x = -2: _____

10. Evaluate the expression $-3\left|x^2 - y\right| - |x|$ when x = -3, y = 10: _____

ALGEBRAIC EXPRESSIONS AND EQUATIONS

CHAPTER 2

- **2.1:** SIMPLIFYING ALGEBRAIC EXPRESSIONS
- **2.2:** SOLVING EQUATIONS
- **2.3:** PROPORTIONS
- **2.4:** CONVERT UNITS OF MEASUREMENT ACROSS MEASUREMENT SYSTEMS
- **2.5:** FORMULAS
- **2.6:** SOLVING INEQUALITIES
- **2.7:** PROBLEM SOLVING

2.1: SIMPLIFYING ALGEBRAIC EXPRESSIONS

simplify algebraic expressions?

Example:

Simplify: $-2x(3y)$

SOLUTION:

$-2x(3y)$
$-2 \cdot 3xy$
$-6xy$

Examples:

Simplify

1. $-5(2x)$

2. $\dfrac{2}{3} \cdot \dfrac{3}{2} x$

3. $-4a(-2b)(-3c)$

4. $(-2he)(2l)(-1p)$

Distributive Property

Evaluate or simplify: $2(3+5)$ $2(3)+2(5)$

Example:

Use distributive property to simplify $-2(x + 3)$

SOLUTION:

$-2(x + 3)$

$-2x - 2(3)$

$-2x - 6$

Examples:

Simplify

1. $2(5x + 3)$

2. $-3(-2 - 4x + 2y)$

3. $-(x - 4)$

4. $4(6x - 5)2$

Like Terms - terms with the same variables raised to the same powers.

Like Terms	Unlike Terms
2x and 5x	2x and 3y
$-4x^3$ and $2x^3$	$5x^2$ and $-6x^3$
$\frac{2}{5}x^2y$ and x^2y	x^2 and x^2y

Examples:

Simplify:

1. $-3x + 5x + 4x^2$

2. $5x - 6 + 8x - 1$

3. $-3x^2y + 5xy^2$

4. $5x - 2x^2 + 3 - 4x^2$

5. $5(x + 3) - 2(4 - x) - (3x - 1)$

Why do we need to simplify algebraic expressions?

_____
_____
_____

2.1: SIMPLIFYING ALGEBRAIC EXPRESSIONS PRACTICE PROBLEMS

Simplify

1. $-3(2x)(-6)$ _____

2. $\left(\dfrac{15}{7}a\right)\left(\dfrac{14}{3}b\right)$ _____

3. $a(5b)(-3c)(-4d)$ _____

4. $-2xy(5.5z)$ _____

5. $2(x+5)$ _____

6. $-4(2x+3y-x)$ _____

7. $-(3a-b-5c)$ _____

9. $-2(x+4y-5z)3$ _____

10. $5a+3a^2-a-a^2$ _____

11. $4xy-5xy^2+3xy$ _____

12. $5-3x(x-4)+3x-x^2$ _____

13. $3(4x+3)-(-x-5)-2(3x+4)$ _____

CHAPTER 2

2.2: SOLVING EQUATIONS

How do you solve equations?

Equation: statement indicating that two expressions are equal

Example:

Is 10 a solution of the equation -3x + 5 = 5x – 25?

Solution:

$$-3x + 5 = 5x - 25$$
$$-3(10) + 5 = 5(10) - 25$$
$$-30 + 5 = 50 - 25$$
$$-25 = 25$$

10 is not a solution of -3x + 5 = 5x – 25.

Examples:

1. Is -2 a solution of the equation $-x + 8 = 3x + 16$?

2. Is -1 a solution of the equation $x^2 + 8 = -x^2 - 8$?

Solving Equations (One Step)

Example: $x - 9 = 12$
Solution: $\begin{aligned} x - 9 &= 12 \\ +9 &+9 \\ x &= 21 \end{aligned}$
Check: $\begin{aligned} x - 9 &= 12 \\ 21 - 9 &= 12 \\ 12 &= 12 \end{aligned}$

Example: $x + 5 = 32$
Solution: $\begin{aligned} x + 5 &= 32 \\ -5 &-5 \\ x &= 27 \end{aligned}$
Check: $\begin{aligned} x + 5 &= 32 \\ 27 + 5 &= 32 \\ 32 &= 32 \end{aligned}$

Examples:

Solve for x

1. $x - 10 = -5$

2. $x + 7 = 1$

3. $-12 + x = 20$

4. $6 + x = -3$

Solving Equations (Two Steps)

Example:	$2x - 3 = 5$
Solution:	$2x - 3 = 5$ $ +3 +3$ $2x = 8$ $\div 2 \div 2$ $x = 4$
Check:	$2x - 3 = 5$ $2(4) - 3 = 5$ $8 - 3 = 5$ $5 = 5$

Example:	$-3x + 5 = -13$
Solution:	$-3x + 5 = -13$ $ -5 -5$ $-3x = -18$ $\div -3 \div -3$ $x = 6$
Check:	$-3x + 5 = -13$ $-3(6) + 5 = -13$ $-18 + 5 = -13$ $-13 = -13$

Examples:

Solve for x

1. $-x + 4 = 10$

2. $\dfrac{x}{2} - 3 = -5$

3. $5 - 6x = 21$

4. $-7 + \dfrac{2}{3}x = 1$

More Solving Equations

Examples:

1. $\dfrac{3}{4}x = 6$

2. $-\dfrac{1}{4}x + 6 = 9$

3. $7 - x = 13$

4. $-5(x-3) + 3x = 11$

Solving Equations with variables on both sides

Example:

$6x - 15 = 4x + 13$

Solution:

$6x - 15 = 4x + 13$

Move variables to one side and numbers to other side of the equality sign

$$6x - 15 = 4x + 13$$
$$ -4x -4x$$
$$2x - 15 = 13$$
$$ +15 +15$$
$$2x = 28$$
$$\div 2 \div 2$$
$$x = 14$$

Check:

$6x - 15 = 4x + 13$
$6(14) - 15 = 4(14) + 13$
$84 - 15 = 56 + 13$
$69 = 69$

Examples:

1. $6x - 7 = 4x + 3$

2. $-7x + 3 = 4x - 19$

3. $-3(4x - 5) = 5x + 3x - 15$

4. $-(x - 4) - 5x = 4(-8 - 3x)$

ALGEBRAIC EXPRESSIONS AND EQUATIONS

Special Cases:

The solution does not have to just be one number; there can be many solutions or no solution.

e. $3(x + 5) - 4(x + 4) = -x - 1$

f. $-4(x - 3) + 2x = 2(10 - x)$

Solving Equations with fractions

Example:

$$\frac{1}{6}x + \frac{5}{2} = \frac{1}{3}$$

SOLUTION:

$$\frac{1}{6}x + \frac{5}{2} = \frac{1}{3}$$

To eliminate the fractions multiply the ENTIRE equation by the LCD. The LCD of 2, 3, and 6 is 6.

$$6\left[\frac{1}{6}x + \frac{5}{2} = \frac{1}{3}\right]$$

$$6\left[\frac{1}{6}x\right] + 6\left[\frac{5}{2}\right] = 6\left[\frac{1}{3}\right]$$

$$x + 15 = 2$$
$$-15 \quad -15$$
$$x = -13$$

Examples:

Solve for x

$$\frac{2}{3} = -\frac{2}{3}x + \frac{3}{4}$$

$$-\frac{5}{6}x - 3x = \frac{1}{3}x + \frac{11}{9}$$

How do we solve equations?

_____

_____

50 CHAPTER 2

2.2: SOLVING EQUATIONS PRACTICE PROBLEMS

1. Is -3 a solution of $-x - 5 = 2x - 2$?

2. Is -2 a solution of $-x^2 + 10 = -3|x| + 12$?

Solve for the variable. Check your work.

3. $-7 + x = -7$ _____

4. $b + (-3) = 8$ _____

5. $-y + 3 = -15$ _____

6. $-5 + \dfrac{a}{4} = 11$ _____

7. $2w + 3 = 10$ _____

8. $3 - 6r = 15$ _____

9. $\dfrac{2}{5}x - 4 = 8$ _____

10. $-2 - \dfrac{3}{4}f = -14$ _____

11. $2(3k - 4) = 2$ _____

12. $-(x - 8) + 3x + 5 = 17$ _____

13. $3x + 4 = 2x - 5$ _____

14. $2(3r + 4) = 5(-2r - 8)$ _____

15. $\dfrac{2}{5}x - 4 = \dfrac{1}{2}$ _____

16. $-2 - \dfrac{1}{2}f = \dfrac{1}{3}f$ _____

17. $x + 4 = -(x + 2)$ _____

18. $2(3r + 1) + 6 = 5(-2r - 8)$ _____

19. $-\dfrac{2}{9} = \dfrac{5}{6}x - \dfrac{1}{3}$ _____

20. $\dfrac{1}{2}b + \dfrac{13}{3} = -\dfrac{3}{4}b + 2b$ _____

2.3: PROPORTIONS

What are proportions and how do you solve them?

Example:

Write a proportion that solves the problem:

A man can eat 5 hamburgers in 2.5 minutes.

How many hamburgers can the man eat in 10 minutes?

a. Set up the proportion:

b. Solve the proportion:

Solution:

a. Set up the proportion:

$$\frac{\text{hamburgers}}{\text{minute}} : \frac{5}{2.5} = \frac{x}{10}$$

b. Solve the proportion:

$$\frac{5}{2.5} = \frac{x}{10}$$

$2.5x = 5(10)$

$2.5x = 50$

$\div 2.5 \quad \div 2.5$

$x = 20$

In 10 minutes the man can eat 20 hamburgers.

ALGEBRAIC EXPRESSIONS AND EQUATIONS

Examples:

Write a proportion that solves the problem:

1. A motorcycle can travel 600 miles on 20 gallons of gasoline.
   How many gallons of gas are needed to travel 100 miles?

   a. Set up the proportion:

   b. Solve the proportion:

2. Jim can type 120 words in 3 minutes.
   How many minutes would it take Jim to type 500 words?

   a. Set up the proportion:

   b. Solve the proportion:

Proportions

Proportion- a mathematical statement that two ratios are equal.

Triangles are similar if their corresponding angles are congruent (equal in measure) or the ratios of their corresponding sides are equal.

Example:

Given two similar triangles:

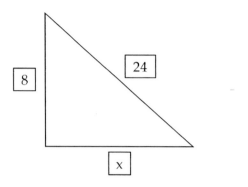 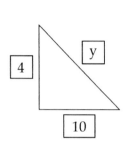

Solve for x using proportions.

Solution:

$$\frac{\text{Large Triangle}}{\text{Small Triangle}} \quad \frac{8}{4} = \frac{x}{10}$$

Solve the proportion by multiplying the diagonals.

$4x = 8(10)$

$4x = 80$

$\div 4 \quad \div 4$

$x = 20$

$$\frac{\text{Large Triangle}}{\text{Small Triangle}} \quad \frac{8}{4} = \frac{20}{10}$$

ALGEBRAIC EXPRESSIONS AND EQUATIONS

Examples:

1. Solve for x using proportions, given two similar triangles:

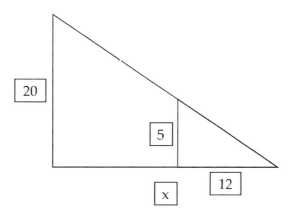

2. Solve for x using proportions, given two similar triangles:

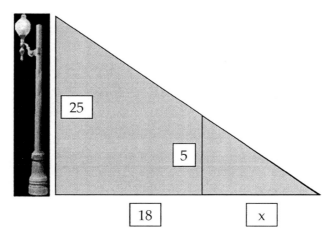

A 25-foot streetlight casts light past a 5 foot tall person, causing a shadow. The person is standing 18 feet from the base of the streetlight. The distance from the man to the edge of the end of the shadow is unknown, "x". How long is the shadow?

What are proportions and how do we solve them?

_____
_____
_____

56 CHAPTER 2

2.3: PROPORTIONS PRACTICE PROBLEMS

1. Given two similar triangles: solve for x using proportions,

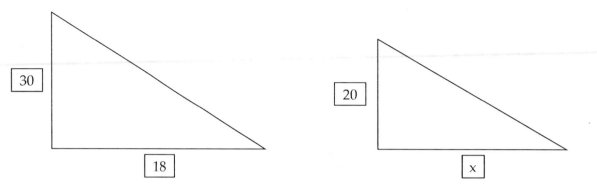

2. Given two similar triangles: solve for x and y using proportions,

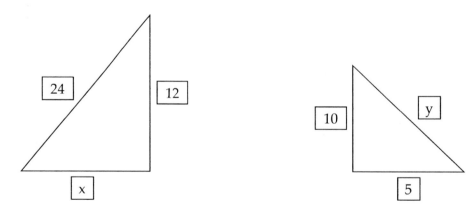

3. Set up a proportion of the following situation (Do not solve the proportion):
   Jim can paint a room in 3 hours. How long would it take Jim to paint 5 rooms?

   Circle which of the following proportions are the same as the proportion of the problem?

   $\dfrac{1}{3} = \dfrac{5}{x}$ $\dfrac{1}{3} = \dfrac{x}{5}$ $\dfrac{x}{5} = \dfrac{3}{1}$ $\dfrac{x}{5} = \dfrac{3}{1}$ $\dfrac{3}{1} = \dfrac{5}{x}$ $\dfrac{5}{1} = \dfrac{x}{3}$

4. Set up and solve the following proportion:
   A car can travel 500 miles on 25 gallons of gas.
   How many miles can the car travel on 10 gallons of gas?

5. Set up and solve the following proportion:
   If 5 pounds of candy cost $6, how much does 7 pounds of candy cost?

2.4: CONVERT UNITS OF MEASUREMENT ACROSS MEASUREMENT SYSTEMS

How does one Convert units of measurement across measurement systems?

Length		
Metric		U.S. System
1 meter (m)	≈	1.094 yards (yd)
1 meter (m)	≈	3.281 feet (ft)
1 kilometer (km)	≈	0.621 miles (mi)
2.540 centimeters (cm)	≈	1 inch (in)
0.305 meters (m)	≈	1 feet (ft)
1.609 kilometer (km)	≈	1 mile (mi)

Capacity		
Metric		U.S. System
1 liter (L)	≈	1.057 quarts (qt)
1 liter (L)	≈	0.264 gallons (gal)
3.785 liters (L)	≈	1 gallon (gal)
0.946 liter (L)	≈	1 quart (qt)
29.574 milliliter (ml)	≈	1 fluid ounce (fl oz)

Weight (Mass)		
Metric		U.S. System
1 kilogram (kg)	≈	2.205 pounds (lb)
0.454 kilograms (kg)	≈	1 pound (lb)
28.350 grams (g)	≈	1 ounce (oz)

Temperature
Celsius (C) to Fahrenheit (F)
$F = \frac{9}{5}c + 32$ $\qquad$ $c = \frac{5}{9}(F - 32)$

ALGEBRAIC EXPRESSIONS AND EQUATIONS

Steps to convert units of measurements across measurement systems

1. Develop a plan for going from the current measurement unit to the desired measure unit.
2. Carry out the plan, making sure you end up with the proper units.

Practice Example:

If a woman is 170 centimeters tall, how many inches tall is she? (Round 2 decimal places)

Step 1:

Develop a plan for going from the current measurement unit to the desired measure unit.

So we refer to the previous chart and find a path from centimeters to inches, and the conversion is:

$$2.54 \text{ centimeters (cm)} \approx 1 \text{ inch (in)}$$

Step 2:

Carry out the plan:

$$\frac{170 \text{ cm}}{1} * \frac{1 \text{ in}}{2.54 \text{ cm}}$$

$$\frac{170 \text{ cm}}{1} * \frac{1 \text{ in}}{2.54 \text{ cm}} = \frac{170}{2.54} in = 66.93 \text{ inches}$$

Practice Example:

If a man weighs 150 pounds, how much does he weigh in kilograms?

Step 1:

Develop a plan for going from the current measurement unit to the desired measure unit.

We are talking about WEIGHT (MASS), so refer to the previous chart and find a path from pounds to kilograms, and the conversion is:

$$1 \text{ kilogram (kg)} \approx 2.205 \text{ pounds (lb)}$$

Step 2:

Carry out the plan:

$$\frac{150 lb}{1} \cdot \frac{1 kg}{2.205 lb}$$

$$\frac{150 lb}{1} \cdot \frac{1 kg}{2.205 lb} = \frac{150}{2.205} kg = 68.027 kg$$

Practice Example:

If a barrel contains 10 gallons of water, how many liters are in the barrel? (Round 1 decimal place)

Step 1:
Develop a plan for going from the current measurement unit to the desired measure unit.

We are talking about CAPACITY, so we refer to the previous chart and find a path from gallons to liters and the conversion are:

One way:	Another way:
3.79 liters (L) ≈ 1 gallon (gal)	1 liter (L) ≈ 0.26 gallons (gal)

Step 2:
Carry out the plan:

One way:	Another way:
3.79 liters (L) ≈ 1 gallon (gal)	1 liter (L) ≈ 0.26 gallons (gal)
$\frac{10 gal}{1} * \frac{3.79}{1\, gal}$	$\frac{10 gal}{1} * \frac{1L}{0.26\, gal}$
$\frac{10\, \cancel{gal}}{1} * \frac{3.79 L}{1\, \cancel{gal}} = \frac{37.9}{1} L = 37.9$ liters	$\frac{10\, \cancel{gal}}{1} * \frac{1L}{0.26\, \cancel{gal}} = \frac{10}{0.26} L = 38.5$ liters

The solutions are close, but they are not exactly the same. Remember when we convert the conversions are approximations so they will be close, but may not be exact.

Practice Example:

If the temperature outside is 90 degrees Fahrenheit, what is the temperature in Celsius?

Step 1:
Develop a plan for going from the current measurement unit to the desire measure unit.

We are talking about TEMPERATURE, so we refer to the previous chart and find a path from Fahrenheit to Celsius and the conversion is:

$$c = \frac{5}{9}(F - 32)$$

Step 2:
Carry out the plan:

$$C = \frac{5}{9}(90 - 32) = 32.22 \text{ Celsius}$$

ALGEBRAIC EXPRESSIONS AND EQUATIONS

PRACTICE:

1. If someone kicked a ball 12 meters, how many feet did the ball go?

2. If Jim drank 2 quarts of water, how many liters did he drink?

3. If you had 1,000 grams of sugar, how many ounces did you have?

4. If the temperature was 37 degrees Celsius, what would the temperature be in Fahrenheit?

How do you convert units of measurement across measurement systems?

_____
_____

2.4: CONVERT UNITS PRACTICE PROBLEMS

(Round answers to 2 decimals places, if necessary):

1. If someone ran a 100 yard dash, how many meters did he run?

2. If I filled a bathtub with 15 gallons, how many liters did I fill?

3. If a pet weighed 3.5 kilograms, how many pounds did it weigh?

4. If the temperature was 80 degrees Fahrenheit, what would the temperature be in Celsius?

ALGEBRAIC EXPRESSIONS AND EQUATIONS 63

2.5:
FORMULAS

Why do we need to solve for variables in formulas?

Formulas used in the real world:

Retail price = Cost + Markup	$r = c + m$
Sale price = Original cost - Discount	$s = c - d$
Interest = (Principal)(Rate)(Time)	$I = PRT$
Perimeter of Rectangle = 2 (Length) + 2 (Width)	$P = 2L + 2W$
Area of a Rectangle = Length(Width)	$A = l \cdot w$
Area of a Triangle = One half(Base)(Height)	$A = \frac{1}{2} \cdot b \cdot h$
Volume of Cylinder = π(Radius)² (Height)	$V = \pi \cdot r^2 \cdot h$

Examples:

1. Find the perimeter of a rectangle given that the length is 12 feet and the width is 10 feet.

3. Find the length of a rectangle given that the perimeter is 100 feet and the width is 15 feet.

Solving Formulas for specific variable:

1. Solve for m: $r = c + m$

2. Solve for R: $I = PRT$

3. Solve for h: $V = \pi \cdot r^2 \cdot h$

ALGEBRAIC EXPRESSIONS AND EQUATIONS

Solving an equation for a specific variable:

Example: Solve for z: $x = -5u - 4z$
Solution: **Step 1**: Isolate z to one side: $x = -5u - 4z$ $\underline{+5u \quad +5u }$ $x + 5u = -4z$
Step 2: Get z alone: $\dfrac{x + 5u}{-4} = \dfrac{-4z}{-4}$ $\dfrac{x + 5u}{-4} = z$
Step 3: Divide both terms by -4 $-\dfrac{x}{4} - \dfrac{5u}{4} = z$
Step 4: Pull the fractions in front: $z = -\dfrac{1}{4}x - \dfrac{5}{4}u$

Examples:

1. Solve for x: $w = 2x + 4z$

2. Solve for b: $6a = -3b + 4c$

Why do we need to solve for variables in formulas?

_____

_____

CHAPTER 2

2.5: FORMULAS PRACTICE PROBLEMS

1. The formula for volume of a rectangular solid is : Volume = Length(Width)(Height) [V=lwh] Find the height, given that the Volume of a rectangle is $300ft^3$, the length is 12 feet, and the width is 5 feet.

2. The formula for the surface area of a rectangular solid is:

   Surface Area = 2(length)(width) + 2(length)(height) + 2(width)(height) or SA = 2lw + 2lh + 2wh

   Find the width, given that the surface area of a rectangle solid is $164m^2$, the length is 4 meters, and the height is 3 meters.

3. The formula to find distance, given the rate and time, is: Distance = rate(time) or D = rt. Solve for "r"

4. The area of a trapezoid is: Area = ½ (Height)(Base 1 + Base 2) or A = ½ h (b1 + b2) Solve for "b1"

5. Solve for x: $3d = x - 2w$

6. Solve for w: $3a = -3wz$

7. Solve for b: $p = 3b + 4v$

8. Solve for y: $2x = -h - 5y$

9. Solve for x: $8y = 8x + 4z$

10. Solve for z: $7x = 2y + 10z$

2.6: SOLVING INEQUALITIES

What is the new feature when solving for inequalities?

Inequality Symbols

<   less than
>   greater than
≤   less than or equal to
≥   greater than or equal to

Notation when using a line graph

○   used for less than or greater than
●   used for less than or equal to or greater than or equal to

Notation when using interval notation

( )   used for less than or greater than
[ ]   used for less than or equal to or greater than or equal to

Example:

Graph $-2 < x \leq 3$. This is read as x is greater than -2 and less than or equal to 3.
Use line graph:

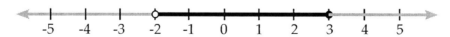

Using interval notation:

$$(-2, 3]$$

ALGEBRAIC EXPRESSIONS AND EQUATIONS 69

Examples:

1. Graph $x \leq -3$

   Use line graph:

   Use interval notation:

2. Graph $x > 3$ or $x < 1$

   Use line graph:

   Use interval notation:

3. Graph $-1 \leq x < 7$

   Use line graph:

   Use interval notation:

Solving Inequalities

Example: Is 10 a solution to $2x - 5 > 7$?

> **REMEMBER:**
>
> If you multiply or divide both sides by a negative, then the inequality changes.

Example:

Solve for x: $2x - 5 > 7$.

SOLUTION:

Solve for x:

$2x - 5 > 7$
$ + 5 +5$
$2x > 12$
$\div 2 \div 2$
$ x > 6$

CHECK:

Choose a number greater than 6 then substitute into original inequality.

Example:

Solve for x: $-3x + 2 > 23$

SOLUTION:

Solve for x:

$-3x + 2 > 23$
$ - 2 -2$
$-3x > 21$
$\div -3 \div -3$
$ x < -7$

CHECK:

Choose a number less than -7, then substitute into original inequality.

Note: Divided by a negative number so the inequality switched form > to <.

ALGEBRAIC EXPRESSIONS AND EQUATIONS 71

Examples:

1. Solve for x: $10 - 2x > 15$

2. Solve for x: $6x - 8 \leq 8x + 10$

3. Solve for y: $8(y + 1) \geq 2(y - 4) + y$

CHAPTER 2

Compound Inequality

1. Solve for x: $-5 \leq x + 4 \leq 6$

2. Solve for x: $-12 < 2x - 6 < 16$

3. Solve for x: $-11 \leq 3(7 - x) < 8$

What is different when you solve for inequalities?

_____

_____

2.6: SOLVING INEQUALITIES PRACTICE PROBLEMS

1. Graph $x \geq -4$
   Use line graph:

   Use interval notation: _____

2. Graph $x < 0$
   Use line graph:

   Use interval notation: _____

3. Graph $-3 < x < 2$
   Use line graph:

   Use interval notation: _____

4. Graph $-4 < x \leq -1$
   Use line graph:

   Use interval notation: _____

Practice Examples:

5. Solve for x: $3x - 4 > 1$

6. Solve for y: $12 - y \geq 11$

7. Solve for x: $2x - 5 \geq 4x + 8$

8. Solve for w: $-2(3w - 4) + 5 < 4(-w + 7)$

9. Solve for z: $-5 \leq z + 5 \leq 12$

10. Solve for x: $-11 < 3 - 4x < 19$

11. Solve for y: $-3 \leq -2(1 - 3y) < 10$

12. Solve for b: $-2 \leq 6(3 - b) + 4b < 25$

2.7: PROBLEM SOLVING

How does one solve word problems?

Problem Solving Strategy
1. Analyze the problem (possibly draw a picture)
2. Form an equation
3. Solve the equation
4. State the conclusion
5. Check the result

Example:

A CD is priced at $15.00, but it is on sale for 20% off. What is the sale price of the CD?

1. **Analyze the problem**

    Given: Original Price: $15.00   **Discount**: 20% off   **Unknown**: x (sale price)

2. **Form an equation**

    Original Price − Discount = Sale Price

    $15 - 0.20(15) = x$

3. **Solve the equation**

    $15 - 0.20(15) = x$
    $15 - 3 = x$
    $x = 12$

4. **State the conclusion**

    The sale price of a CD that was $15.00 discounted at 20% is $12.00

5. **Check the result**

    $15 - 0.20(15) = x$
    $15 - 0.20(15) = 12$
    $15 - 3 = 12$
    $12 = 12$

Problem Solving with solutions afterwards

Problem solving with fractions, decimals, and percents

1. A pair of jeans is priced at $50.00, but is on sale for 25% off. What is the sale price of the pair of jeans?

2. If a television costs $311 after a 35% discount, what was the original cost?

3. If a DVD player costs $306 after a 30% increase in price, what was the original cost?

Problems with simple interest: $(I = P \cdot R \cdot T)$ Interest = (Principal)(Rate)(Time)

4. Find the amount of money necessary to be invested now at 5% simple interest to yield $200 accumulation in 8 years.

ALGEBRAIC EXPRESSIONS AND EQUATIONS

Problem solving with Measurement and Area

5. The length of a rectangular garden is 8 meters more than its width. Its perimeter is 76 meters. Find the length of the garden.

6. The perimeter of a triangle is 30 inches. The length of the middle side is 2 inches more than the length of the smaller side and the largest side is 4 inches more than twice the length of the smallest side. Find the length of the smallest side.

Problem solving with translation

7. If 8 times a number is increased by 20, the result is 26 less than the square of the number. Translate this statement where a number is represented by "x".

Problem solving with proportions

8. Create a proportion that solves the problem: A car can travel 603 miles for 11 gallons of gasoline. How far can the car travel for 36 gallons?

Problem solving with the formula: Distance = (Rate)(Time)

9. Two cars start from the same point and travel in opposite directions. The rate of the slower car is 15 miles per hour less than the rate of the faster car. After 8 hours they are 840 miles apart. Find the speed of the cars.

   r = faster
   $r - 15$ = slower
   $t = 8$
   $D = 840$

Answers to Problems Solving

1. To find the discount you multiply the original price by the percent of the discount.

   So, the discount is $50.00(0.25) = $12.50.

   To find the sale price you subtract the original price from the discount.

   So, to find the sale price you take 50.00 − 12.50 = 37.50. The sale price is $37.50.

2. Original Cost − Discount = Sale Price

   $X - .35X = 311$

   $1X - .35X = 311$

   $$\frac{.65X}{.65} = \frac{311}{.65}$$

   $X = 478.46$

   Check: $478.46 - .35(478.46) = 311$

   $478.46 - 167.46 = 311$

   $311 = 311$

3. Original Cost + Increase = New Price

   $X + 0.30X = 306$

   $1X + 0.30X = 306$

   $$\frac{1.30X}{1.30} = \frac{306}{1.30}$$

   $X = \$235.39$

   Check: $235.39 + .30(235.39) = 306$

   $235.39 + 70.61 = 306$

   $306 = 306$

4. The formula to find simple interest is: Interest = Principal x Rate x Time (I = P*R*T)

   The information given leads to: 200 = X (0.05) (8)

   Then simplify: 200 = X(0.4)

   Then solve for x by dividing both sides by 0.4: 200/0.04 = X(0.04)/0.04

   X = $500.00, the principal (amount to be invested) is $500.00.

   Check: 200 = 500(0.05)(8)

   200 = 200

5.

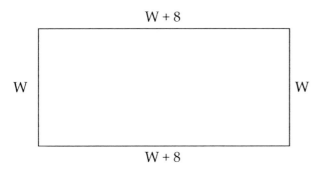

Width: W
Length: W + 8
Perimeter: 76

Equation: W + W + 8 + W + W + 8 = 76   Width: W = 15

Combine like terms: 4W + 16 = 76   Length: 15 + 8 = 23

$$\frac{-16 \quad -16}{\frac{4W}{4} = \frac{60}{4}}$$

Check: 15 + 23 + 15 + 23 = 76

76 = 76

Solve: W = 15

6.

Small: X
Middle: X + 2
Large: 2X + 4
Perimeter: 30

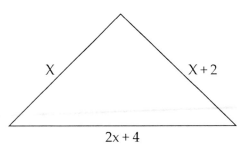

The equation is: X + X + 2 + 2X + 4 = 30
Combine like terms: 4X + 6 = 30
Solve:
$$\begin{array}{r} -6 \quad -6 \\ \frac{4X}{4} = \frac{24}{4} \\ X = 6 \end{array}$$

Check: 6 + 6 + 2 + 2(6) + 4 = 30
6 + 8 + 16 = 30
14 + 16 = 30
30 = 30

Small: X = 6
Middle: 6 + 2 = 8
Large: 2(6) + 4 = 16

___

7. Translate the problem to:
$8X + 20 = X^2 - 26$

___

8. $\dfrac{miles}{gallons} = \dfrac{603}{11} = \dfrac{x}{36}$

Note: If your proportion does not match the proportions shown, multiply the diagonals to see if it matches.

Another way: $\dfrac{11}{603} = \dfrac{36}{x}$

9.

Type	D	R	T
Fast Car	8X	X	8
Slow Car	8(X – 15)	X – 15	8
Total	840		

Fill in the 4 x 4 chart with the given information.

To fill in the distance column use the formula Distance = Rate x Time

You now have the formula:

$8X + 8(X - 15) = 840$

Distributive Property: $8X + 8X - 120 = 840$

Combine like terms: $16X - 120 = 840$

Solve for X:
$$+120 \quad +120$$
$$\frac{16X}{16} = \frac{960}{16}$$
$$X = 60$$

Substitute X into the chart and answer the question:

Type	D	R	T
Fast Car	8(60) = 480 miles	60 mph	8
Slow Car	8(60 – 15) = 360 miles	60 – 15 = 45 mph	8
Total	840 miles		

_____
_____
_____
_____
_____

What is our strategy for problem solving?

_____
_____
_____
_____
_____

Additional Problem Solving

Problem solving with fractions, decimals, and percents

1. The average length of an American alligator is 11 feet. Suppose this average has a margin of error of 5% more than 11 feet, or as little as 5% less than 11. Using 5% as the margin of error, what is the minimum and the maximum average length of an American alligator?

2. After dinner two friends, Jim and Nancy, received the check for their meal. Jim is paying $\frac{2}{3}$ of the check and leaving no tip. Nancy is paying the remainder of the check. Nancy is leaving 30% of the check as a tip. Using X as the amount of the check, create an algebraic expression for the check amount for Jim and another algebraic expression for the check amount for Nancy.

3. Given that there are 300 million people living in the United States and 45 million of them are left handed, what percent of people are left handed?

4. Aiko paid $75 for a concert ticket. She must also pay a 9% convenience fee for purchasing the ticket online. What is the total cost of Aiko's concert ticket including the convenience fee?

Problem solving with simple interest

The formula for simple interest is: Interest = (Principal) (Rate) (Time)

1. In the third year, after investing $8,000 in the bank, a man received a check for $720 in interest. What interest rate did his money have during the 3 years? Assume simple interest.

2. Frank needs to borrow $15,000 to purchase a used car. The bank is offering him a 3.9% simple interest loan and is charging $3,510 in interest during the life of the loan. How long is the life of the loan?

3. The bank is offering 4.5% interest for 4 years and Angelo wants to earn $360 in interest at the end of the 4 years. How much would Angelo need to invest to earn the $360 interest at the interest rate of 4.5% for 4 years? Assume simple interest.

Problem solving with Measurement and Area

1. A rectangular lot is 42 feet wide. To put fencing around the yard for the dog the home owner would need 154 feet of fencing? How many feet is the length of the yard?

2. Large tiles are 18 inches by 18 inches. If the floor is 17,172 square inches, how many tiles do you need for your floor?

3. Carpet sells for $3 per square foot and will cost you $390 to recarpet a room. If the rectangular room is 10 feet long, how many feet wide is it?

4. A triangular table top is being made of wood. Find the number of square centimeters of wood that is needed for the triangular table top that has a base of 123 centimeters and a height of 92 centimeters.

ALGEBRAIC EXPRESSIONS AND EQUATIONS

2.7: PROBLEM SOLVING PRACTICE PROBLEMS

1. A CD is priced at $15.00, but it is on sale for 20% off. What is the sale price of the CD?

2. If a sony play station costs $250 after a 15% discount, what was the original cost?

3. If a palm pilot costs $1300 after a 20% increase in price, what was the original cost?

4. Find the simple interest percent if you invested $1000.00 for 5 years and you received $500.00 in interest.

5. The width of a rectangular garden is 8 meters less than its length. Its perimeter is 76 meters. Find the length of the garden.

6. The perimeter of a triangle is 51 inches. The length of the middle side is 5 inches more than the length of the smaller side and the largest side is 4 inches less than three times the length of the smallest side. Find the length of the middle side.

7. If 10 times a number is decreased by 29, the result is the product of 42 and a number. Write an equation to represent this statement where a number is represented with "x".

8. Create a proportion that solves the problem: A car can travel 1200 miles on 60 gallons of gasoline. How many gallons do you need to travel 100 miles?

9. Two shrimp boats start from the same port at the same time, but they head in opposite directions. The slower boat travels 15 knots per hour slower than the fast boat. At the end of 12 hours, they were 600 nautical miles apart. How many nautical miles had the slow boat traveled by the end of the 12-hour period?

10. If a student had difficulty completing problem solving questions, what should the student do?
    a. Ask questions
    b. Always go to class and take good notes
    c. Complete their homework
    d. Seek extra help when needed (instructor/tutoring)
    e. All the above!!!

MAT0028C DEVELOPMENTAL MATH II - TEST 1 (CHAPTERS 1 / 2) REVIEW

QUESTIONS FROM CHAPTERS 1 AND 2

1. Add: (-7) + 5 + (-3)

2. Subtract: 5.8 – (-2.8)

3. Multiply: (-7)(2)(-3)

4. Solve: $\dfrac{2}{3} + \dfrac{1}{4}$

5. Solve: $4\dfrac{1}{2} - 2\dfrac{3}{16}$

6. Solve: $\left(3\dfrac{2}{7}\right)\left(1\dfrac{1}{13}\right)$

7. Solve: $\left(-2\dfrac{1}{4}\right) \div \left(1\dfrac{1}{13}\right)$

8. Solve: 40 = -8(x - 3)

9. Solve: $\frac{3}{5}y - 4 = 2$

10. Graph: $-18 < 3x - 6 < 9$

11. Simplify: $6 - 18 \div 9 - 4$

12. Simplify: $23 - (8)2 \div (14 - 6) \cdot 6$

13. Simplify: $|8 + (-14)| + 9$

14. Evaluate $-6w^2 + 5w + 3$ when $w = -4$:

15. Solve for y: $\frac{6}{5}y - \frac{2}{3} = 4$

16. Solve for t: $x = -8z + 7t$

ALGEBRAIC EXPRESSIONS AND EQUATIONS

17. Five less than twice the square of a number, the result is 7 more than three times a number. Write the equation that could be used to find this number, x.
    DO NOT SOLVE THE PROBLEM, JUST SET UP THE EQUATION.

18. Write a proportion that solves the problem: A motorcycle can travel 705 miles on 19 gallons of gasoline. How many gallons of gas are needed to travel 1253 miles?
    DO NOT SOLVE THE PROBLEM, JUST SET UP THE PROPORTION.

19. Solve the inequality: $14x + 4 \leq 26x + 20$

20. If someone ran 200 meter dash, how many yards is that?

21. If a container filled with 5 liter, how many gallons is that?

22. If a person weighed 100 pounds, how many kilograms is that?

23. If the temperature was 90 degrees fahrenheit, what is the temperature in celsius?

24. A jacket is priced at $75.00, but is on sale for 20% off. What is the sale price of the jacket?

25. If a digital camera costs $375 after a 25% discount, what was the original cost?

26. If a DVD player costs $690 after a 15% increase in price, what was the original cost?

27. Find the simple interest percent to yield $100 interest in 5 years when $500 is invested.

28. The length of a rectangular pool is 15 less than three times its width. Its perimeter is 50 meters. Find the length of the pool.

GRAPHING

3 CHAPTER

3.1: GRAPHING USING THE RECTANGULAR COORDINATE SYSTEM

3.2: GRAPHING LINEAR EQUATIONS USING TABLE OF VALUES

3.3: GRAPHING LINEAR EQUATIONS USING INTERCEPTS

3.4: SLOPE OF A LINE

3.5: SLOPE INTERCEPT

3.1: GRAPHING USING THE RECTANGULAR COORDINATE SYSTEM

Why do we need to know how to graph using the rectangular coordinate system?

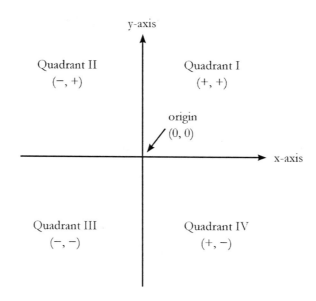

Key points:

Axes- two perpendicular lines used to locate points on a plane

Origin- (0, 0); where the 2 axes intersect

Quadrants- 4 regions that the axes divide the plane

x-axis- horizontal axis

y-axis- vertical axis

Ordered pair- (x-coordinate (abscissa), y-coordinate (ordinate))

Notice how the quadrants are numbered starting at the top right and move counter clockwise.

Exercise:

Graph and label the points

(-3, 4), (3, 4), (-4, 0), (-2, -2), (0, -2.5), (2, -2), (4, 0)

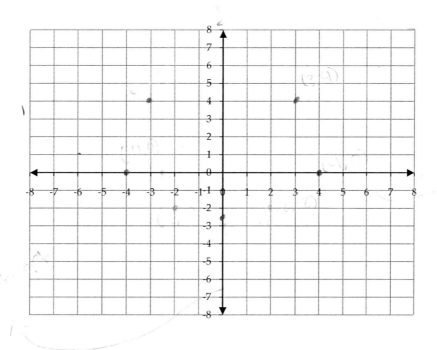

Label the coordinates for each letter

A: _____   J: _____   N: _____   O: _____

R: _____   S: _____   T: _____

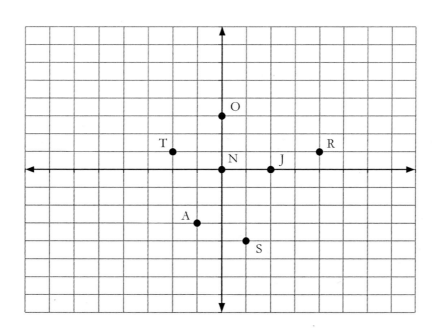

Graph Mathematical Relationships

The time to fill a tub of water follows the table below.

Examine the relationship between the time in minutes and the water in the tub in gallons and determine what the missing values are. Then use the graph below to label the axes, label the scales, and plot the points.

Time (minutes)	Water in tub (gallons)
0	0
1	8
3	24
4	32

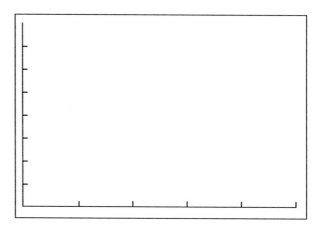

Looking at the plotted points, what do you notice about the relationship between the time and the amount of water in the tub?

_____

_____

Reading Graphs

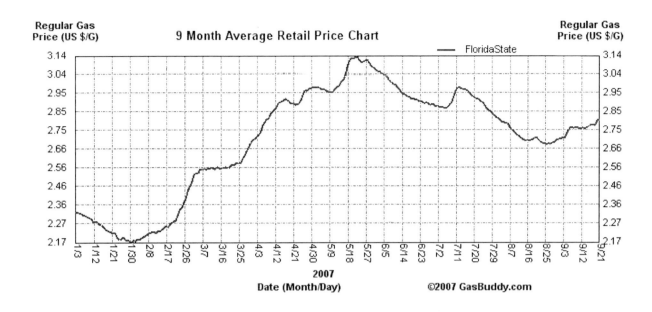

The x-axis represents time from January 3, 2007 to September 21, 2007.

The y-axis represents the average price of gas in Florida per gallon.

How much did gas cost in Florida on January 12, 2007? _____

How much did gas cost in Florida on March 16, 2007? _____

How much did gas cost in Florida on September 12, 2007? _____

When was the gas price in Florida the least amount per gallon and how much was it?

When: _____

How much: _____

When was the gas price in Florida the most amount per gallon and how much was it?

When: _____

How much: _____

When was the gas price $2.75 per gallon? _____

Why do we need to know how to graph using the rectangular coordinate system?

_____

_____

96 CHAPTER 3

3.1 GRAPHING USING THE RECTANGULAR COORDINATE SYSTEM PRACTICE PROBLEMS

1. Plot and label the following points:

    (0, 0), (1, 2), (2, -4), (-3, -3), (-4, 5.5), (0, -5), (1, 0), (6, 0), (-4.5, 0)

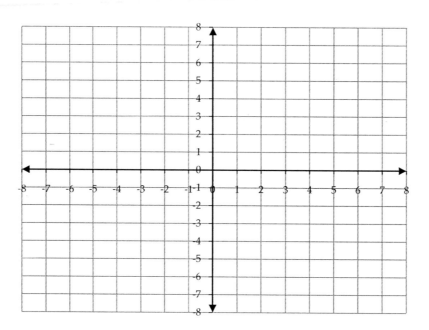

2. Review the table and examine the relationship. Fill in the missing values and plot the points. Label the axes and label the scales.

Year	Price of video game
2000	60
2001	50
2003	30
2005	10

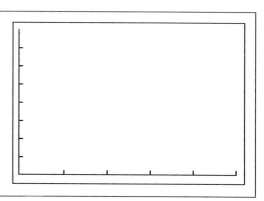

What is the relationship between the year and the price of video game?

_____

_____

3. Examine the graph and answer the following questions.

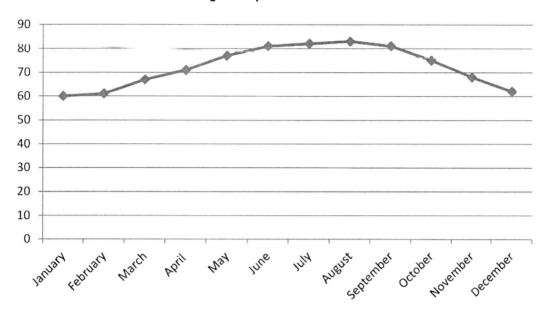

What is the average temperature in Florida in January? _____

What is the average temperature in Florida in June? _____

What is the average temperature in Florida in October? _____

What month has the lowest average temperature in Florida and what is that temperature?

Month: _____

Temperature: _____

What month has the highest average temperature in Florida and what is that temperature?

Month: _____

Month: _____

When is the average temperature 80 degrees? _____

3.2: GRAPHING LINEAR EQUATIONS USING TABLE OF VALUES

How to graph linear equations?

Solutions of Equations in Two Variables

Example: Determine whether each ordered pair is a solution of: $2x - y = 10$

a. (2, -6) b (10, -10)

Example: Complete the table of solutions for: $y = 2x + 3$

x	y	(x, y)
-2		
-1		
0		
1		
2		
3		

Graph the points from the example: $y = 2x + 3$

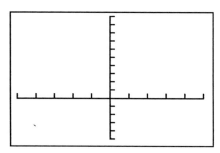

Example: Complete the table of solutions for: $2x - 3y = 12$

x	y	(x, y)
0		
	0	
3		
	2	
-3		
	-1	

Graph the points from the example: $2x - 3y = 12$

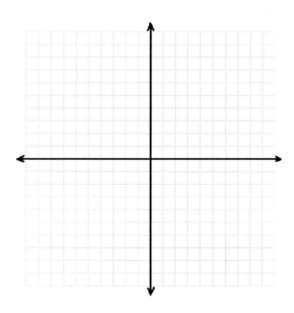

Graphing Linear Equations

A linear equation is a first degree (the variable is raised only to the first power) polynomial equation; its graph is a straight line in the Cartesian coordinate system.

Linear Equations are equations that can be written as: $Ax + By = C$ (standard form) where A, B, and C are real numbers and A, B are not 0 at the same time.

A linear equation can be written in other forms such as slope-intercept form ($y = mx+b$), this will be discussed in depth in a following section.

Graph: y = 2x + 4

Step 1:

Find at least 3 solutions, using a table of values

x	y	(x, y)

Step 2:

Plot the solutions and connect the coordinates

Graph using a table of values: y = 3x – 2

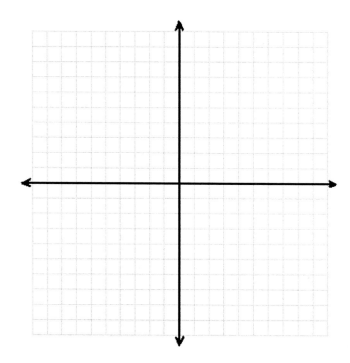

x	y	(x, y)

GRAPHING 101

Graph using a table of values: 2x – 4y = 6

x	y	(x, y)

What is one way to graph linear equations?

_____

_____

3.2 GRAPHING LINEAR EQUATIONS USING TABLE OF VALUES PRACTICE PROBLEMS

1. Determine whether each ordered pair is a solution of: $x - 4y = 12$
   a. (0, -3)   b. (-4, 4)

2. Complete the table of solutions for: $y = -x + 5$
   Then plot the coordinates.

x	y	(x, y)
-2		
-1		
0		
1		
2		
3		

3. Complete the table of solutions for: $3x - 2y = 6$
   Then plot the coordinates.

x	y	(x, y)
0		
	0	
4		
	9	
-3		
	-4	

GRAPHING 103

4. Graph using a table of values: $y = 2x - 1$

x	y	(x, y)

5. Graph using a table of values: $5x - 2y = 10$

x	y	(x, y)

6. Write a linear equation for this situation, then graph it.

The cost to rent a car is $30.00 and $0.50 per mile, where the total cost is "y" and the amount of miles used is "x".

_____

Graph the linear equation using a table of values. Make sure you label x axis, y axis, and scale.

Hint: Use realistic values for x, number of miles.

x	y	(x, y)

3.3:
GRAPHING LINEAR EQUATIONS USING INTERCEPTS

What is another way to graph linear equations?

Linear Equation- equations that can be written as Ax + By = C (standard form)

There are other forms of linear equations such as: Slope-intercept form: $y = mx + b$. This will be explained later in section 3.5.

Intercepts

x –intercept -the point where the graph crosses the x –axis

Find the x-intercept by making y = 0.

y –intercept -the point where the graph crosses the y –axis

Find the y-intercept by making x = 0.

Graph the given equation: $2x - 3y = 6$

Find the x-intercept: _____

x	y

Find the y-intercept: _____

x	y

Graph the equation

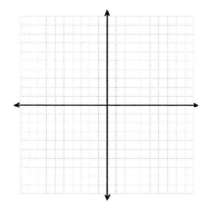

CHAPTER 3

Examples:

1. Graph the given equation using intercepts: $3x = -5y + 8$

   Find the x-intercept: _____   Find the y-intercept: _____

x	y

x	y

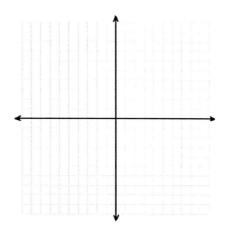

2. Graph the given equation using intercepts: $y = -3x + 6$

   Find the x-intercept: _____   Find the y-intercept: _____

x	y

x	y

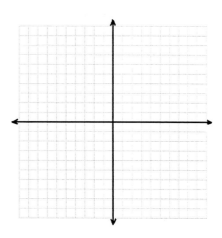

What is another way to graph linear equations?

_____

3.3 GRAPHING LINEAR EQUATIONS USING INTERCEPTS PRACTICE PROBLEMS

1. Graph the given equation using intercepts: $3x - 4y = 12$.

   Find the x intercept: _____   Find the y-intercept: _____

x	y

x	y

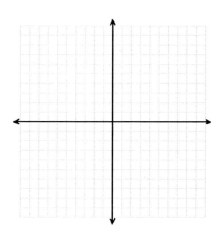

2. Graph the given equation using intercepts: $y = 4x - 5$

   Find the x-intercept: _____   Find the y-intercept: _____

x	y

x	y

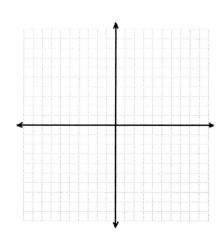

3. Graph the given equation using intercepts:
y = 2x + 3

Find the x-intercept: _____
Find the y-intercept: _____

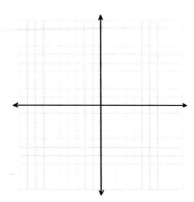

4. Graph the given equation using intercepts:
3x = y - 7

Find the x-intercept: _____
Find the y-intercept: _____

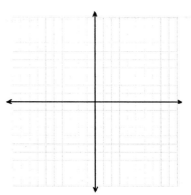

5. Graph the given equation using intercepts:

$$y = \frac{1}{2}x - 4$$

Find the x-intercept: _____
Find the y-intercept: _____

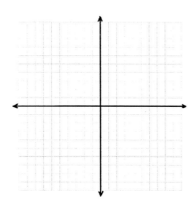

6. Graph the given equation using intercepts:

$$\frac{1}{4}x + \frac{2}{3}y = 2$$

Find the x-intercept: _____
Find the y-intercept: _____

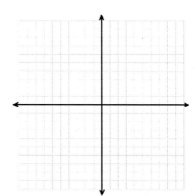

3.4:
SLOPE OF A LINE

What is the slope of a line?

Slope- the ratio $\dfrac{\text{change in } y - \text{coordinate}}{\text{change in } x - \text{coordinate}}$ as we move from one point to another.

$$\dfrac{Rise}{Run} : \begin{array}{ll} + = Up & - = Down \\ - = Left & + = Right \end{array}$$

Example:

Below is a picture of a line, $y = -\dfrac{4}{3}x + \dfrac{1}{3}$ and two points on the line, (-2, 3) and (1, -1) that are highlighted. Given the two points on the line, find the slope.

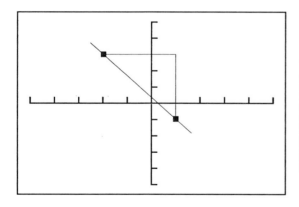

Starting from (-2, 3) & moving to (1, -1), the slope is:

$$\dfrac{Rise}{Run} = \dfrac{-4}{+3} = -\dfrac{4}{3}$$

Starting from (1, −1) & moving to (−2, 3), the slope is the same:

$$\dfrac{Rise}{Run} = \dfrac{+4}{-3} = -\dfrac{4}{3}$$

Find the slope of this example:

Below is a picture of a line, $y = \dfrac{3}{5}x - \dfrac{1}{5}$ and two points on the line, (-3, -2) and (2, 1) that are highlighted. Given the two points on the line, find the slope.

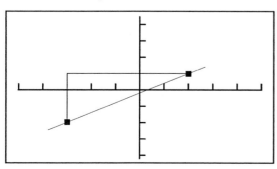

Starting from (−3, −2) moving to (2, 1) the slope is:

$$\dfrac{Rise}{Run} = \underline{} = \underline{}$$

Starting from (2, 1) moving to (−3, −2) the slope is:

$$\dfrac{Rise}{Run} = \underline{} = \underline{}$$

Examples of Slopes:

Positive Slopes:

Negative Slopes:

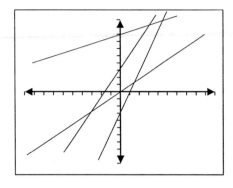
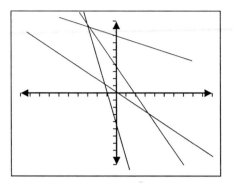

Examples:

Find the slope of the following:

m = _____

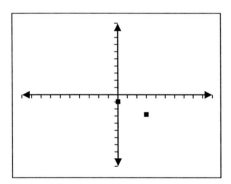

m = _____

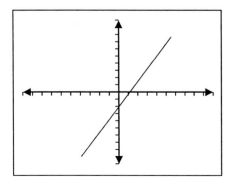

what does slope mean?

Slope - the rate of change of the steepness of the line

Example:

Al would like to improve his health by joining a gym. To join a gym there is a one time membership fee of $100 and a monthly fee of $30.

Write an equation to describe this situation, using "y" as the total cost and "x" as the number of months going to the gym.

The equation for this situation is:

_____

What part of the situation is changing when x, the amount of months, changes?
Is it the one time membership fee or the total charge of the monthly fees?

_____

To find the slope of this situation:

Label the variables: $\dfrac{\text{Change in y - coordinate}}{\text{Change in x - coordinate}} = \dfrac{\text{Change in Total Cost}}{\text{Change in Total Months}}$

Slope : $\dfrac{Rise}{Run}$:

What does the slope mean in this case?

_____

_____

Find the slope given two points

To find slope for two points we use the slope formula.

Slope formula: $m = \dfrac{y_2 - y_1}{x_2 - x_1}$

Example: Find the slope using the points (6, -3) and (4, 3):

Step 1: Label the points

(6, -3)   (4, 3)

(x_1, y_1) (x_2, y_2)

Step 2: Substitute values into the slope formula and solve.

$$m = \frac{y_2 - y_1}{x_2 - x_1} = \frac{3 - (-3)}{4 - 6} = \frac{6}{-2} = -3$$

Answer: The slope is -3

Example:
Given the points (3, 6) and (5, 2), find the slope:

$$m = \frac{y_2 - y_1}{x_2 - x_1}$$

Example:
Given the points (-2, -2) and (-12, -8), find the slope:

$$m = \frac{y_2 - y_1}{x_2 - x_1}$$

Example:
Given the points (8, -4) and (8, -3), find the slope:

$$m = \frac{y_2 - y_1}{x_2 - x_1}$$

Horizontal and Vertical Lines

Horizontal Lines

The equation of a horizontal line is: $y = k$ (k is a constant)

The slope of a horizontal line is zero.

In standard form it is: $(0)x + by = C$

In slope-intercept form it is: $y = 0x + b$

Example: Graph $y = 2$

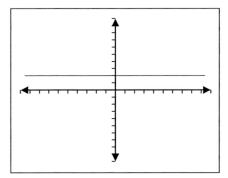

Example: Graph $y = -6$

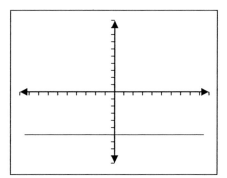

Calculated:

Choose 2 points on line

(1, 2) (-2, 2)

$m = \dfrac{2-2}{-2-1} = \dfrac{0}{-3} = 0$

$m = 0$

Calculated:

Choose 2 points on line

(2, -6) (-3, -6)

$m = \dfrac{-6-(-6)}{-3-2} = \dfrac{0}{-5} = 0$

$m = 0$

Practice: Graph $y = -4$

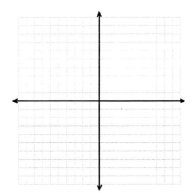

Practice: Graph $y = 3$

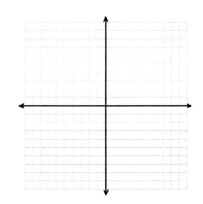

Practice: Graph $y = 0$

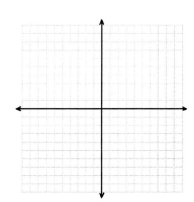

Vertical Lines

The equation of a vertical line is: $x = k$ (k is a constant)

The slope of a vertical line is undefined.

In <u>standard form</u> it is: $Ax + (0)y = C$

In <u>slope intercept</u> form it is: $0 = mx + b$

Example: Graph $x = 5$

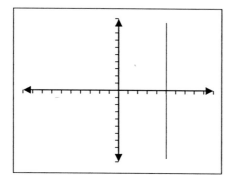

Example: Graph $x = -7$

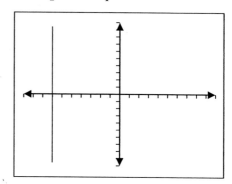

<u>Calculated:</u>

Choose 2 points on line

(5, 2) (5, -3)

$$m = \frac{-3-2}{5-5} = \frac{-5}{0} = \text{undefined}$$

m = undefined

<u>Calculated:</u>

Choose 2 points on line

(-7, 1) (-7, 5)

$$m = \frac{5-1}{-7-(-7)} = \frac{4}{0} = \text{undefined}$$

m = undefined

Practice: Graph $x = -2$

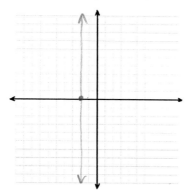

Practice: Graph $x = 4$

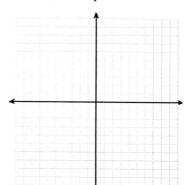

Practice: Graph $x = 0$

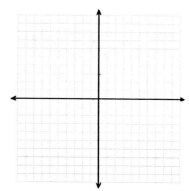

What is the definition of the slope of a line?

_____

_____

3.4 THE SLOPE OF A LINE PRACTICE PROBLEMS

1. Given the picture find the slope:

   m = _____

   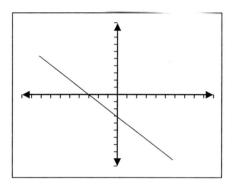

2. Given the picture find the slope:

   m = _____

   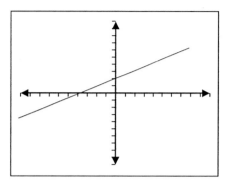

3. Given the picture find the slope:

   m = _____

   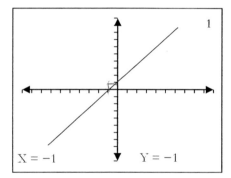

4. Given the picture find the slope:

   m = _____

   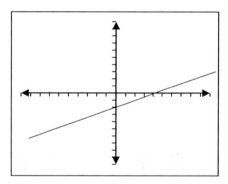

5. To join an online video game website you need to pay a one time registration fee of $20 and $1.50 per hour to use the website.

   What is the equation for the cost of joining the website, where "y" is the total cost and "x" is the number of hours on the website?

   _____

   What is the slope in this situation?

   _____

   Explain the slope in the context of the situation?

   _____

   _____

6. Find the slope given the points (3, 7) and (2, 10)

$$m = \frac{y_2 - y_1}{x_2 - x_1}$$

7. Find the slope given the points (3, -5) and (-6, -7)

$$m = \frac{y_2 - y_1}{x_2 - x_1}$$

8. Find the slope given the points $\left(\frac{1}{2}, -4\right)$ and $\left(\frac{2}{3}, 6\right)$

$$m = \frac{y_2 - y_1}{x_2 - x_1}$$

9. Graph x = 3

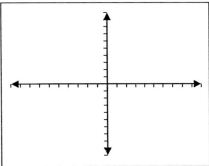

10. Graph x = 0

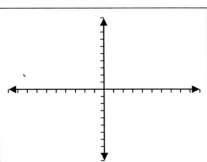

11. Graph y = -4

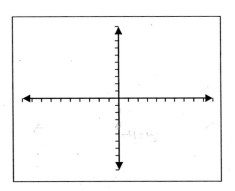

12. Graph y = 5

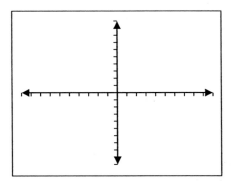

3.5: SLOPE INTERCEPT

Slope-Intercept Form: $y = mx + b$

m = slope b = y-intercept

An example of graphing using slope-intercept form: Graph: $y = -\frac{2}{3}x - 1$

<u>Step 1:</u> Determine the "m" (slope) and "b" (y–intercept)

The equation can be written as: $y = -\frac{2}{3}x + (-1)$ to clearly see "m" and "b".

The m (slope) is: $\frac{-2}{3}$ The b (y-intercept) is: -1

<u>Step 2:</u> First plot your "b" (y-intercept) on the coordinate system. b = -1

Note: A y-intercept is when $x = 0$, so in this example;

$$y = -\frac{2}{3}(0) - 1$$
$$y = -1$$

y-intercept is (0, -1)

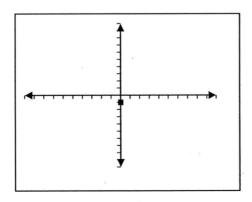

Step 3: Use the "m" (slope) to find another point. The slope is $\frac{-2}{3}$; which means from the y intercept, you should go down 2 and to the right 3.

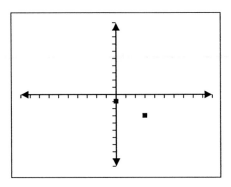

Step 4: Connect the points to graph: $y = -\frac{2}{3}x - 1$

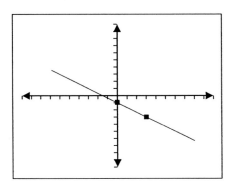

Practice Graphing Linear Equations using the Slope-Intercept Form

Graph the equation $y = -4x - 3$

m = _____   b = _____

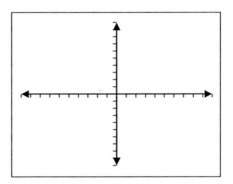

Graph the equation $-3x + 4y = 8$
First step solve the equation for "y"

m = _____   b = _____

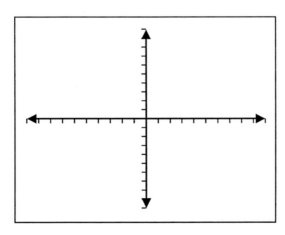

For each of the lines shown below, use your knowledge of slope and intercept to find the equation of the line given. Give your answer in the form: $y = mx + b$.

Each tick mark represents 1 unit.

1. y =

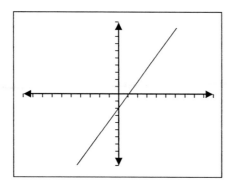

2. y =

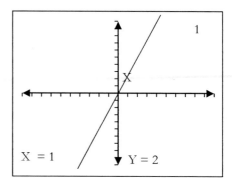

3. y =

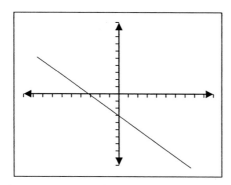

4. y =

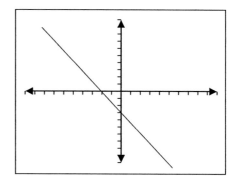

5. y =

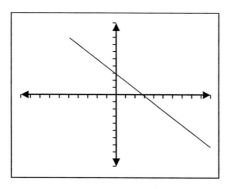

6. y =

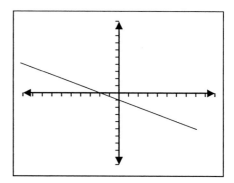

7. y =

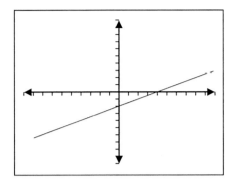

8. y =

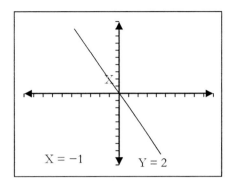

X = −1   Y = 2

9. y =

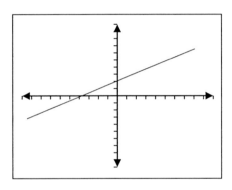

10. y =

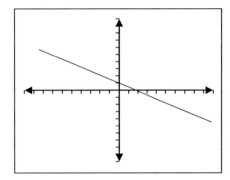

Practice Graphing Linear Equations using the Slope-Intercept Form

Graph the equation $y = -\frac{1}{4}x + 2$.

m = _____   b = _____

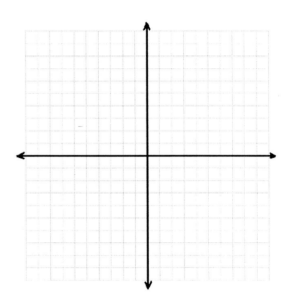

Graph the equation $-2x + 3y = 6$.
First step solve the equation for "y"

m = _____   b = _____

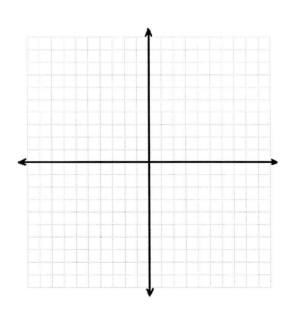

TECHNOLOGY EXTRA: Graphing Using the Calculator (TI 82/83/84)

Graph: $y = -\dfrac{2}{3}x - 1$

<u>Step 1</u>: Press the y – button on the top left of the calculator.

The screen should look like this:

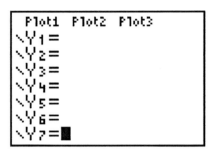

<u>Step 2</u>: After the $y_1 =$ type in the equation.

*Be careful when you type in the fractions and negatives.

It is recommended to put in the fractions in parentheses.

There are different buttons for subtraction sign – and negative sign (-).

Here is how you should type the equation in the calculator: (-2/3)x-1

This is how the screen should look:

<u>Step 3</u>: Then press *graph* button to graph this equation.

Your graph should look like this in a standard window.

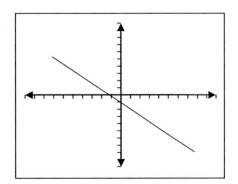

Equation of a Line Using a Real Life Application "Going to the Movies"

Al is inviting his friends to go to the movies on Friday night, but he does not know how many of them are coming and what the total cost will be. He knows that he has to pay $5.00 for parking at Pointe Orlando. Also, the cost of the movies is $10.00 per person.

Write a linear equation for the following example using "x" to represent the number of friends going to the movies and "y" representing the total cost.

Answer the following questions for this real life word problem

1. How much will it cost if Al invites 1 friend to the movies?
   (Do not count Al; this is just the price for the friend.) _____
   After you find the solution, plot the solution (x, y) on the following graph.

2. How much will cost if Al invites three friends to the movies?
   (Do not count Al, this is just the price for the friends.) _____
   After you find the solution, plot the solution (x, y) on the following graph.

3. If Al decided to pay for the trip to the movies for his friends, how many people can Al pay with $55.00? _____
   After you find the solution, plot the solution (x, y) on the following graph.

4. If Al decided to pay for the trip to the movies for his friends, how many people can Al pay with $75.00? _____
   After you find the solution, plot the solution (x, y) on the following graph.

Graph

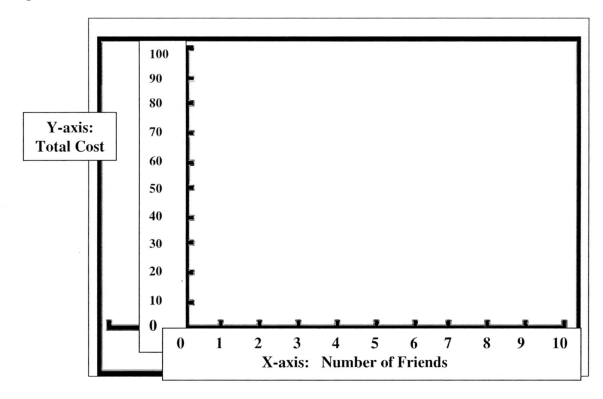

5. What do you notice about the points that were plotted?

_____

6. Is there a relationship between the number of friends (x) and the total cost (y) to go to the movies? If so, what is the relationship?

   Hint: Look at the cost as you add one more friend.

   _____

   _____

7. What is the cost if no one goes to the movies and why?

   _____

3.5: SLOPE INTERCEPT FORM PRACTICE PROBLEMS

1. Given the equation $y = 4x - 3$

   What is the slope (m)? _____

   What is the y-intercept (b)? _____

2. Given the equation $y = -\dfrac{2}{5}x + 6$

   What is the slope (m)? _____

   What is the y-intercept (b)? _____

3. Given the equation $2x - 3y = 12$

   What is the slope (m)? _____

   What is the y-intercept (b)? _____

4. Given the equation $4x + 7y = 14$

   What is the slope (m)? _____

   What is the y-intercept (b)? _____

5. Graph using slope-intercept form

   $y = -x + 4$

   m = _____   b = _____

   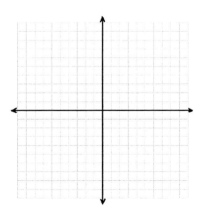

6. Graph using slope-intercept form

   $y = -\dfrac{2}{3}x - 4$

   m = _____   b = _____

   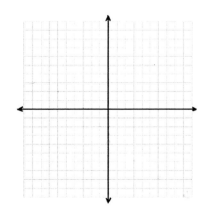

7. Graph using slope-intercept form
$x + 3y = -9$

m = _____ b = _____

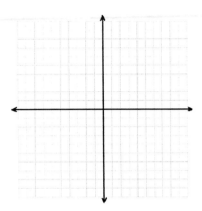

8. Graph using slope-intercept form
$-5x - 4y = -12$

m = _____ b = _____

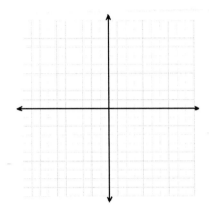

9. Find the equation of the line in slope-intercept form of the graph below:

_____

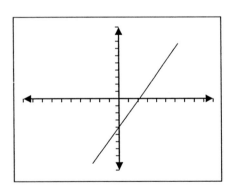

10. Find the equation of the line in slope-intercept form of the graph below:

_____

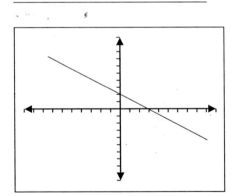

GRAPHING 129

11. You are trying to raise money to buy a new cell phone. You already have $50.00 saved up and you are working at a job where you make $10 per hour.

   a. Write an equation in slope-intercept form that expresses this relationship where "x" is the number of hours you work and "y" is the total amount you have earned?

   _____

   b. What is the slope? _____

   What does the slope mean in the context of the problem?

   _____

   c. What is the y-intercept? _____

   What does the y-intercept mean in the context of the problem?

   _____

   d. How much money would you have saved up after 8 hours of work? _____

   e. How much money would you have saved up after 16 hours of work? _____

   f. How many hours do you have to work if the cost of the cell phone is $300? _____

POLYNOMIALS

4 CHAPTER

4.1: PROPERTIES OF EXPONENTS

4.2: ZERO AND NEGATIVE EXPONENTS

4.3: SCIENTIFIC NOTATION

4.4: POLYNOMIALS

4.5: ADDING/ SUBTRACTING POLYNOMIALS

4.6: MULTIPLYING POLYNOMIALS

4.7: DIVISION OF POLYNOMIALS (MONOMIALS)

4.1: PROPERTIES OF EXPONENTS

2^4 2 is the base and 4 is the exponent

$2^4 = 2 \cdot 2 \cdot 2 \cdot 2 = 16$

Example:

Simplify: $2^3 \cdot 2^5$

Solution

$2^3 \cdot 2^5$
$2 \cdot 2 \cdot 2 \cdot 2 \cdot 2 \cdot 2 \cdot 2 \cdot 2$
2^8

Example:

Simplify: $3^2 \cdot 3^8$

When multiplying exponential expressions with the same base, what is a shortcut to simplify the expression?

_____

_____

Properties of Exponents

$x^a \cdot x^b = x^{a+b}$

Examples:

1. $(-5)^6(-5)^8$

2. $x^5 x^3$

3. $x^7 y^2 x^6 y$

4. $(4a^7 b^3)(-3a^2 b)$

5. $(x-4)^3 (x-4)^5$

Example:

Simplify: $\dfrac{3^5}{3^3}$

Solution:

$$\dfrac{3^5}{3^3}$$

$$\dfrac{3 \cdot 3 \cdot 3 \cdot 3 \cdot 3}{3 \cdot 3 \cdot 3}$$

$$\dfrac{\cancel{3} \cdot \cancel{3} \cdot \cancel{3} \cdot 3 \cdot 3}{\cancel{3} \cdot \cancel{3} \cdot \cancel{3}}$$

$$3^2$$

Example:

Simplify: $\dfrac{4^9}{4^6}$

When dividing exponential expressions with the same base, what is a shortcut to simplify the expression?

_____

_____

Properties of Exponents

$$\frac{x^a}{x^b} = x^{a-b}$$

Examples:

1. $\dfrac{(-3)^7}{(-3)^3}$

2. $\dfrac{x^5}{x^2}$

3. $\dfrac{10x^{12}y^9z^4}{2x^8yz^4}$

4. $\dfrac{-6a^6b^5}{9a^4b^5c^4c^2}$

CHAPTER 4

Example:

Simplify: $(5^3)^4$

Solution:

$(5^3)^4$
$(5^3)(5^3)(5^3)(5^3)$
$(5 \cdot 5 \cdot 5)(5 \cdot 5 \cdot 5)(5 \cdot 5 \cdot 5)(5 \cdot 5 \cdot 5)$
5^{12}

Example:

Simplify: $(2^5)^3$

When raising exponential expressions to a power, what is a shortcut to simplify the expression?

Properties of Exponents

$(x^a)^b = x^{a \cdot b}$

Examples:

1. $(10^5)^{12}$

2. $(x^3)^7$

3. $(2x^3y^4z)^3$

4. $(-3a^2b^8)^4$

5. $\left(\dfrac{3a^5b^2}{2c^4}\right)^3$

4.1: RULES FOR EXPONENTS PRACTICE PROBLEMS

1. $3^4 \cdot 3^7$

2. $(-2)^5 (-2)^2$

3. $a^2 a^3 a$

4. $x^2 x^8 x^7 y^2$

5. $ab^3 c^5 a^{10} c^2 bd^7$

6. $(4)^2 (-6)^3 (4)^6 (-6)$

7. $(-6x^2 y^4)(4x^6 y)$

8. $(5x^2 - 3y)^4 (5x^2 - 3y)^2$

9. $\theta^3 \psi^4 \alpha \theta^5 \psi^2$

10. $\dfrac{7^8}{7^3}$

11. $\dfrac{3^5 (-8)^3}{3^3 (-8)}$

12. $\dfrac{a^{12}}{a^4}$

13. $\dfrac{16b^{10}c^9d}{8b^5cd}$

14. $\dfrac{-2x^7x^3y^5}{4x^4y^3yz^2}$

15. $\left(\dfrac{a^3b^4}{c}\right)\left(\dfrac{a^5c^3}{b}\right)$

16. $(2^3)^4$

17. $(x^5)^6$

18. $(3x^2)^3$

19. $(4x^7y^5z)^2$

20. $(-2a^3bc^5)^5$

21. $\left(\dfrac{5a^2b^3}{c^4d^5}\right)^2$

22. $\left(\dfrac{12a^5a^2}{4a^6b}\right)^3$

23. $\left(\dfrac{-4x^3y^4x^5}{12x^4x^2y}\right)^2$

140 CHAPTER 4

4.2: ZERO AND NEGATIVE EXPONENTS

Example:

Simplify: $\dfrac{2^4}{2^4}$

SOLUTION:

$$\dfrac{2^4}{2^4} = 2^{4-4} = 2^0$$

What is a base raised to 0?

$$\dfrac{2^4}{2^4}$$

$$\dfrac{2 \cdot 2 \cdot 2 \cdot 2}{2 \cdot 2 \cdot 2 \cdot 2}$$

$$\dfrac{\cancel{2} \cdot \cancel{2} \cdot \cancel{2} \cdot \cancel{2}}{\cancel{2} \cdot \cancel{2} \cdot \cancel{2} \cdot \cancel{2}} = 1$$

$2^0 = 1$

Example:

Simplify: $(-5)^0$

When raising an exponential expression to zero, what is the simplified expression (except for 0 and infinity)?

_____

_____

Properties of Exponents

$b^0 = 1$

Examples:

1. $(x^3)^0$

2. $\left(\dfrac{5x^{100}y^{20}z^{151}}{a^{15}b}\right)^0$

3. $x^0 + 2$

4. $3x^0$

5. $(2x^5 y^0)^3$

6. $(5a^5b^3)^0 (4a^6b^0)$

Example:

Simplify: $\dfrac{3^2}{3^6}$

SOLUTION:

$$\dfrac{3^2}{3^6} = 3^{2-6} = 3^{-4}$$

What can we do with a negative exponent?

$$\dfrac{3^2}{3^6}$$

$$\dfrac{3 \cdot 3}{3 \cdot 3 \cdot 3 \cdot 3 \cdot 3 \cdot 3}$$

$$\dfrac{\cancel{3} \cdot \cancel{3}}{\cancel{3} \cdot \cancel{3} \cdot 3 \cdot 3 \cdot 3 \cdot 3}$$

$$\dfrac{1}{3^4}$$

Therefore,

$$3^{-4} = \dfrac{1}{3^4}$$

Example:

Simplify: $\dfrac{2^3}{2^9}$

How do we make a negative exponent positive?

Properties of Exponents

$b^{-1} = \dfrac{1}{b}$

Examples:

1. 6^{-3}

2. $\dfrac{1}{7^{-2}}$

3. $\dfrac{x^{-2}y^5}{z^{-1}}$

4. $\dfrac{-3x^5 y^{-2} z^{-3}}{x^{-2} y^{-4} z^4}$

5. $(-4)^{-1}$

6. $\left(\dfrac{3x^{-2} y^0}{z^5} \right)^{-2}$

4.2: ZERO AND NEGATIVE EXPONENTS PRACTICE PROBLEMS

1. 5^0

2. b^0

3. $5^0 + b^0$

4. $\left(\dfrac{-121a^{55}b^{38}c^{45}}{247x^{17}y^{68}z}\right)^0$

5. $(-2)^0$

6. $(5a^2b^0c^3)^2$

7. $(2x^0y^5)^4 (3x^7y^0)^0$

8. $\dfrac{6^2}{6^3}$

9. $\dfrac{x^7}{x^{10}}$

10. 3^{-2}

11. x^{-5}

12. $\dfrac{1}{5^{-2}a^{-3}}$

13. $\dfrac{a^{-4}b^{-2}}{c^{-3}}$

14. $\dfrac{x^{-3}y^7z^{-1}}{-2a^6b^{-3}d}$

15. $(3x^5 y^{-3})(2x^{-3} y^2)$

16. $(2)^{-3}$

17. $(4x^3 y^{-4})^{-2}$

18. $\left(\dfrac{12x^5 y^{-3} z^{-1}}{4x^{12} y^6 z^{-9}}\right)^{-4}$

4.3: SCIENTIFIC NOTATION

Why do we use scientific notation?

Scientists often deal with extremely large and small numbers.
For example, the distance of the earth to the sun is 150,000,000 kilometers (93,750,000 miles).
Another example is the diameter of the influenza virus, which is 0.00000256 inches.

Standard Form: 2,500,000	**Scientific Notation:** 2.5×10^6
Standard Form: 0.000123	**Scientific Notation:** 1.23×10^{-4}
Scientific Notation: -3.25×10^5	**Standard Form:** $-325{,}000$
Scientific Notation: 5×10^{-7}	**Standard Form:** 0.0000005

Examples:

1. To convert 1.95×10^{-7} to standard notation, start at the decimal point; the -7 means move the decimal point 7 movements to the left (due to the negative). A positive exponent means move the decimal to the right, a negative means move to the left.

$$1.95 \times 10^{-7}$$
$$0.000000195$$

2. To convert 0.000000195 to scientific notation, you should start at the decimal point and count the number of movements to get to the position of after the first non zero number. In this case, the number is 1. You would need to move the decimal 7 movements to the right. Since the original number is less than 1, we will use -7. The number written in scientific notation is:

$$1.95 \times 10^{-7}$$

Examples:

Convert to scientific notation:

1. 3,650,000,000 _____

2. 0.0093 _____

3. −0.000000004 _____

Convert to standard notation:

4. 4.56×10^4 _____

5. 1.2×10^{-5} _____

6. -5×10^{-2} _____

Why do we use scientific notation?

_____

_____

4.3: SCIENTIFIC NOTATION PRACTICE PROBLEMS

Convert to scientific notation:

1. 2,500,000

2. -0.000004653

3. 0.0012

4. −5,000,000,000

Convert to standard notation:

5. 1.3×10^6

6. -6.25×10^{-5}

7. -3.27×10^{10}

8. 3×10^{-7}

4.4: POLYNOMIALS

What is a polynomial?

Degree of a term: the value of the exponent on the variable. If a polynomial is in more than one variable, the degree of a term is the sum of the exponents on the variables. The degree of a nonzero constant is 0.

Degree of a polynomial: the highest degree of any term of the polynomial.

Coefficient: a numerical or constant quantity placed before and multiplying the variable in an algebraic term.

Monomial- algebraic expression containing only one term.
    Examples: $x^2 y$, -4, $3a^2 b^3 c^7 d$, $6x$

Binomial- polynomial with 2 terms.
    Examples: $5x^2 y + 3x^5$, $x - 4$, $5a^2 b^3 c^7 d + 6x^2 y^5 z^3$, $3x^2 - 5x$

Trinomial- polynomial with 3 terms.
    Examples: $4x^3 y^2 - 5x^2 y + 3x^5$, $x^2 + 2x - 4$, $9a^2 b^3 + 8x^2 y^5 - 7$

Polynomial- a mathematical expression written as a sum of terms.
    Examples: $5x^2 y + 3x - 4y + 2$, $10x^2 - 4$, $5a^2 b^3 c^7 d$, $3x^2 - 4x + 6$

    Polynomial with more than 3 terms - Polynomial with _____ terms.

    Example: Polynomial with 4 terms: $x^3 - 5x^2 + 3x - 1$
    Example: Polynomial with 5 terms: $5x^4 y - 8x^2 y^3 + 6x - 15y + 3$

Classify as MONOMIAL, BINOMIAL, or TRINOMIAL:

$x^2 + 2x + 1$ _____

$5x^3 - 4x^2$ _____

$12a^2 b^3 c^7 dx^2 y^5 z^3$ _____

$6x^3 - 5x^3$ _____

Examples:

1. Given the polynomial: $4x^7 - 2x^4 y^2 - 6x^2 y + 3y^3$

    List the terms: _____

    List the coefficients of each term: _____

    Find the degree of each term: _____

    Find the degree of the polynomial: _____

2. Given the polynomial: $ab + 6a^3 b^2 - 4ab^7 + 8$

    List the terms: _____

    List the coefficients of each term: _____

    Find the degree of each term: _____

    Find the degree of the polynomial: _____

Evaluate a polynomial:

1. $-2x^2 + x - 5$ when $x = -1$

2. $(x-5)^2 - x^2$ when $x = -2$

Graphing nonlinear equations: Use the table of values

1. Graph: $y = x^2 - 3$

x	y

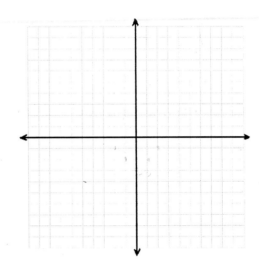

2. Graph: $y = x^3 + 2$

x	y

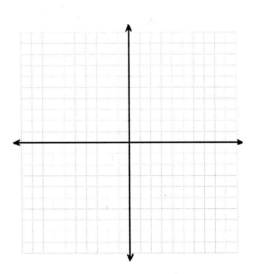

What is a polynomial?

_____

_____

4.4: POLYNOMIALS PRACTICE PROBLEMS

1. What types of polynomials are these? Make sure they are simplified.
   Choices: monomial, binomial, trinomial, polynomial with ____ terms

   a. $4x^2y^5z + 2x^7y^6$ _____

   b. x^2 _____

   c. $3a^2b + 4a^2b - a^2b - 7a^2b$ _____

   d. $5x^2 - 3x + 6$ _____

   e. $x^3 - 2x^2 + 3x + 6$ _____

2. Given the polynomial: $3x^2 - 5x + 7$

   a. List the terms: _____

   b. List the coefficients of each term: _____

   c. Find the degree of each term: _____

   d. Find the degree of the polynomial: _____

3. Given the polynomial: $a^6 + 9a^5b^3 - 2ab - 8b^3 + 3$

   a. List the terms: _____

   b. List the coefficients of each term: _____

   c. Find the degree of each term: _____

   d. Find the degree of the polynomial: _____

154 CHAPTER 4

Evaluate questions 4 to 7

4. $x^2 + 2x + 1$ when $x = -1$

5. $-3x^2 - x + 7$ when $x = -2$

6. $-2xy - x^3$ when $x = -3$, $y = -2$

7. $x^3 - y^2 - 4x + y$ when $x = 2$, $y = -1$

8. Graph: $y = x^2 + x - 6$

x	y

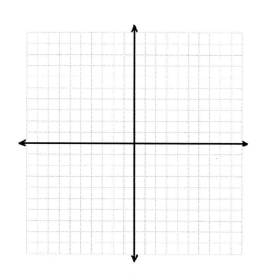

4.5: ADDING/SUBTRACTING POLYNOMIALS

How do we add and subtract polynomials?

Simplify:
Add
$(6x^2 - 2x - 5) + (x^2 - 3x + 4)$

Solution:
$(6x^2 - 2x - 5) + (x^2 - 3x + 4)$

Distribute the addition
$6x^2 - 2x - 5 + x^2 + -3x + 4$

Combine like terms
$6x^2 + x^2 - 2x + -3x - 5 + 4$

Simplified solution:
$7x^2 - 5x - 1$

Simplify:
Subtract
$(3x^2 - 4x + 6) - (6x^2 - 7)$

Solution:
$(3x^2 - 4x + 6) - (6x^2 - 7)$

Distribute the subtraction
$3x^2 - 4x + 6 - 6x^2 - (-7)$

Combine like terms
$3x^2 - 6x^2 - 4x + 6 - (-7)$

Simplified solution:
$-3x^2 - 4x + 13$

Examples:

Simplify:

1. $5x^2 - 3x + 4x^2 - 7x + 6$

2. $2xy - 7y + 3y - 2xy - 4xy^2$

3. $(x^2 - 5x - 7) + (3x^2 + 4x - 5)$

4. $(-4x + 6) - (2x^2 - 6x^2 - 3)$

5. $3(-2x + 5) - 5(3x^2 - 4x - 1)$

6. $x(9x^2 - x + 5) - 2x(4x^2 + 3x - 6)$

How do we add and subtract polynomials?

4.5: ADDING/ SUBTRACTING POLYNOMIALS PRACTICE PROBLEMS

Simplify:

1. $4 - 7x^2 - 2x^3 + 5x^3 - 3x - 7x^2 + 3x + 9$

2. $a^2b - 7ab^2 + 5ab - 8ab^2 - 4a^2b + 3a$

3. $(x^2 - 5) + (-6x^2 + 2x - 5)$

4. $(-a^2 + 9a - 1) + (a^2 + 9a - 8)$

5. $(5x^2 - 3x + 2) - (2x^2 - 7x - 2)$

6. $(-x^2 - 4x) - (x^2 + 6x - 2)$

Simplify:

7. $(-8x + 7) + (5x - 3) + (-2x - 7)$

8. $-3x^2 - (x - 7) - (x^2 + 6x - 1)$

9. $(-2x + 5) + (3x^2 - 4x - 8) - (x^2 + 5x - 3)$

10. $4x^2y + (6x^2y - 2xy^2) - (x^2y + 10xy^2)$

11. $(x^2 + 3) - (x - 2)$

12. $3x(6x^2 + 2) - x(x^2 + 7x - 5)$

4.6: MULTIPLYING POLYNOMIALS

How do we multiply polynomials?

Simplify: $2x^4(3x^2 - 4x + 5)$

Solution:
Distributive Property
$2x^4(3x^2 - 4x + 5)$
$2x^4(3x^2) + 2x^4(-4x) + 2x^4(5)$
$6x^6 - 8x^5 + 10x^4$

Examples:

1. $-3x^2(-x^4 + 3x^2 - 1)$

2. $-5x^3y(3x^2y^4 - 7x^2 + 4y^3 - 2)$

Simplify: $(3x + 2)(5x - 9)$

Solution:

Method 1: FOIL

$(3x + 2)(5x - 9)$

Use FOIL (First, Outer, Inner, Last)

First : $(3x)(5x) = 15x^2$

Outer : $(3x)(-9) = -27x$

Inner : $(2)(5x) = 10x$

Last : $(2)(-9) = -18$

Combine like terms

$15x^2 - 27x + 10x - 18$

Simplified solution :

$15x^2 - 17x - 18$

Method 2: Table Method

$(3x + 2)(5x - 9)$

	$5x$	-9
$3x$	$15x^2$	$-27x$
$+2$	$10x$	-18

Combine like terms

$15x^2 - 27x + 10x - 18$

Simplified solution:

$15x^2 - 17x - 18$

Examples:

1. $(x-5)(-3x-8)$

   Optional Table Method

2. $(4x+1)(x^2-6x-8)$

   Optional Table Method

3. $(2x^2-4x+1)(7x^2-3x-4)$

   Optional Table Method

Special Products

Examples:

1. Simplify:

    $(x+3)^2$

    Optional Table Method

2. Simplify:

    $(2x-5)^2$

    Optional Table Method

3. Simplify:

   $(x+3)(x-3)$

   Optional Table Method

4. Simplify:

   $(4x-1)(4x+1)$

   Optional Table Method

How do we multiply polynomials?

_____

_____

4.6: MULTIPLYING POLYNOMIALS PRACTICE PROBLEMS

Simplify:

1. $-x^3(-3x^5 + 15x^3 - 9x - 2)$

2. $-3a^2b^4(a^3b^3 - 2a^5b + 4a^5b^2c^3 - 1)$

3. $(x+5)(x+3)$

4. $(x-7)(x+9)$

5. $(2x+7)(x-6)$

6. $(4x-5)(3x-2)$

7. $(-3x+4)(2x+1)$

8. $-(x+3)(2x-9)$

Simplify:

9. $(x + 2)(x^2 - 3x + 4)$

10. $(3x - 2)(2x^2 - 5x - 7)$

11. $(x^2 + 2x + 1)(x^2 - 3x + 2)$

12. $(2x^2 - 3x - 4)(3x^2 - 5x + 1)$

13. $(x + 4)^2$

14. $(3x - 7)^2$

15. $(x + 5)(x - 5)$

16. $(5x - 6)(5x + 6)$

4.7: DIVISION OF POLYNOMIALS (MONOMIALS)

How do you divide polynomials (monomials)?

Example:

Simplify: $\dfrac{30x^4y^3}{15x^2y^7}$

SOLUTION:

$$\dfrac{30x^4y^3}{15x^2y^7}$$

$$\dfrac{2\cdot 3\cdot 5xxxxyyy}{3\cdot 5xxyyyyyyy}$$

$$\dfrac{2\cdot \cancel{3}\cdot \cancel{5}\cancel{x}\cancel{x}xxyyy}{\cancel{3}\cdot \cancel{5}\cancel{x}\cancel{x}yyyyyyy}$$

$$\dfrac{2x^2}{y^4}$$

Examples:

1. $\dfrac{6x^5y^2z^2}{20x^5yz^{10}}$

2. $\dfrac{-3ab^5d^0}{9a^6b^2c^3}$

Example:

Simplify: $\dfrac{10x^2 - 8x + 4}{2x}$

Solution:

$$\dfrac{10x^2 - 8x + 4}{2x}$$

$$\dfrac{10x^2}{2x} - \dfrac{8x}{2x} + \dfrac{4}{2x}$$

$$5x - 4 + \dfrac{2}{x}$$

Examples:

1. $\dfrac{12x^4 - 3x}{6x^3}$

2. $\dfrac{20x^3y^3 - 7xy^5 + 35x^6}{5x^2y^3}$

How do we you divide polynomials (monomials)?

_____

_____

4.7: DIVISION OF POLYNOMIALS (MONOMIALS) PRACTICE PROBLEMS

Simplify:

1. $\dfrac{30a^3b^5}{5a^4b^5}$

2. $\dfrac{-12x^{12}y^8z^0}{-24x^{16}y^{10}z^5}$

3. $\dfrac{20x-25}{5}$

4. $\dfrac{14x^2-21x}{7x}$

5. $\dfrac{x^5-6x}{3x^3}$

6. $\dfrac{9b^5-3b^4+18}{9b^2}$

7. $\dfrac{6a^2b^7-3a^7b^{10}+5a^2b^0}{3a^5b^3}$

8. $\dfrac{12x^6y-15x^5y^5+2x^4}{6x^4y}$

MAT0028C DEVELOPMENTAL MATH II   Name: _____

TEST 2 (CHAPTERS 3 / 4) REVIEW

Questions from Chapter 3

1. Plot the following points and label
   A (-2, 3)   B (0, 5)
   C (-6, 0)   D (0, 0)
   E (-7, -1)

2. Is (-2, 5) a solution to $-x + y = 3$?

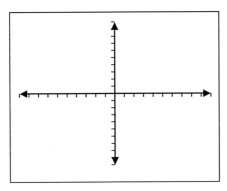

3. Complete the table for the equation:
   $3x - y = 6$

4. Plot the points in #3

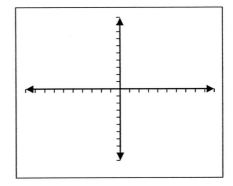

x	y	(x, y)
-2		
-1		
0		
	6	
	-3	
	3	

5. Find the x-intercept of: $y = -3x + 6$

6. Find the y-intercept of: $2x - 3y = 12$

7. Find the slope of: (-3, 2) and (-5, -4)

$$m = \frac{y_2 - y_1}{x_2 - x_1}$$

8. Graph: $y = -\frac{2}{5}x - 4$

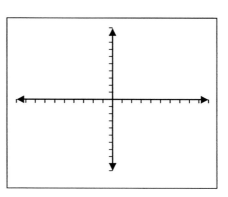

9. What is the equation of the line for:

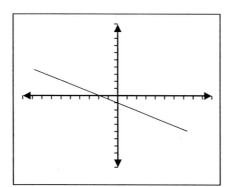

10. Graph: $y = -3x + 2$

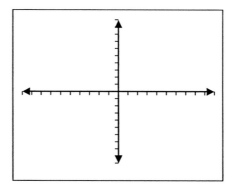

11. Simplify: $x^5 y^3 x^6 y$

12. Simplify: $\dfrac{10x^5 y^7 z^2}{2x^8 y z^0}$

13. Simplify: $(2x^2 y^6)^4 x^3 y^7$

14. Simplify: $\dfrac{x^{-3} y^5 z^{-2}}{x^0 y^{-1} z^4}$

15. Simplify: $(x^3 y^0)^{-5}$

16. Simplify: $\left(\dfrac{3x^{-3} y^{-2} z^7 w^0}{9 y^{-5} z^{-6}}\right)^{-2}$

17. Convert to scientific notation: 1,530,000

18. Convert to standard notation: -2.5×10^{-3}

19. Simplify: $(5x^2 - 3x + 4) - (-2x^2 + 6x - 5)$

20. Simplify: $-3x^2(2x^4 - 5x^3)$

21. Simplify: $(2x - 3)(-x + 8)$

22. Simplify: $(4x - 5)^2$

23. Simplify: $\dfrac{10x^5y^3 - 2xy^2}{2xy^2}$

24. Simplify: $\dfrac{4x^3y^5 - 8x + 2y^4}{4x^2y^4}$

Review Questions

25. Simplify: $-2[7(3x - 4) + 5x]$

26. Evaluate $-2w^2 + 4w - 6$ when $w = -1$:

27. Solve for x: $-4(-5x + 2) = 2(2x + 20)$

28. Solve for y: $\dfrac{2}{3}y - \dfrac{3}{4} = 2$

29. Solve for b: $c = 4a + 2b$

30. If a digital player costs $425 after a 15% discount, what was the original cost?

31. TRANSLATE: 8 less than the square of a number, the result is the product of 6 and a number. Create the equation that could be used to find this number, x.
    DO NOT SOLVE THE PROBLEM, JUST SET UP THE EQUATION.

32. Write a proportion that solves the problem: A hybrid can travel 1100 miles on 55 gallons of gasoline. How many gallons of gas are needed to travel 1925 miles?
    DO NOT SOLVE THE PROBLEM, JUST SET UP THE PROPORTION.

FACTORING/ RATIONAL EXPRESSION

CHAPTER 5

- **5.1:** FACTORING BY GCF AND GROUPING
- **5.2:** FACTORING TRINOMIALS IN THE FORM OF $x^2 + bx + c$
- **5.3:** FACTORING TRINOMIALS IN THE FORM OF $ax^2 + bx + c$
- **5.4:** FACTORING DIFFERENCE OF TWO SQUARES AND PERFECT SQUARE TRINOMIALS
- **5.5:** FACTORING USING MULTIPLE METHODS
- **5.6:** SOLVE QUADRATIC EQUATIONS BY FACTORING
- **5.7:** FACTORING APPLICATIONS
- **5.8:** SIMPLIFY RATIONAL EXPRESSIONS
- **5.9:** MULTIPLY AND DIVIDE RATIONAL EXPRESSIONS
- **5.10:** ADD AND SUBTRACT RATIONAL EXPRESSIONS WITH MONOMIAL DENOMINATORS

5.1: FACTORING BY GCF AND GROUPING

What is factoring?

Find the prime factors of:

1. 6
2. 27
3. 250

4. $25x^2y^5$
5. $100ab^3$
6. $45x^4y^2z$

The greatest common factor (GCF)- the largest common factor of the integers.

To Find the GCF:
1. Find the prime factorization of the terms.
2. Find the common factors in each of the terms and the most of each in all terms.
3. Multiply the most common factors in all of the terms.

FACTORING/RATIONAL EXPRESSION

Example:

Find the GCF of: 36 and 90

Step 1:
Find the prime factorization of the terms

```
         36                    90
       6 · 6                 9 · 10
     2·3·2·3               3·3·2·5

      2·3·2·3               2·3·3·5
```

Step 2:
Find the common factors in each of the terms and the most of each in all terms

The common factors are: 2, 3

The most 2's in common is: 2

The most 3's in common is: 3 · 3

Step 3:
Multiply the most common factors in all of the terms.

The GCF is: 2·3·3 = 18

Examples:

1. Find the GCF of: 24   and   70

2. Find the GCF of: 45   60   75

CHAPTER 5

Example:

Find the GCF of: $9x^3 y^2$ and $15xy^4$

Step 1:

Find the prime factorization of the terms

$$9x^3 y^2 \qquad\qquad 15xy^4$$

$$3 \cdot 3xxxyy \qquad\qquad 3 \cdot 5xyyyy$$

Step 2:

Find the common factors in each of the terms and the most of each in all terms

The common factors are: 3,x,y

The most 3's in common is: 3

The most x's in common is: x

The most y's in common is: yy

Step 3:

Multiply the most common factors in all of the terms.

The GCF is: $3xyy = 3xy^2$

Examples:

1. Find the GCF of: $\qquad 20x^5 y^6 \qquad$ and $\qquad 150x^7 y^3$

2. Find the GCF of: $\qquad 42a^3b^2 \qquad 63ab^5 \qquad 21a^5b^4$

Simplify: $3x(2x + 5)$

Now work backwards.

Factor using Greatest Common Factor

Example:

Factor $-25x^3y^3 + 15x^5y^6 - 5x^2y^7$

If we break down the polynomial into prime factors we would have:

$5 \cdot 5xxxyyy + 5 \cdot 3xxxxxyyyyyy - 5xxyyyyyyy$

What are the most factors you can pull out of all of the terms?

$5x^2y^3$

Then what remains once the GCF is pulled out?

Divide each term by the GCF

$$\frac{25x^3y^3}{5x^2y^3} + \frac{15x^5y^6}{5x^2y^3} - \frac{5x^2y^7}{5x^2y^3} =$$
$$5x \quad + 3x^3y^3 \quad - \quad y^4$$

The factored form is: $5x^2y^3(5x + 3x^3y^3 - y^4)$

You can check by multiplying.

Examples:

1. Factor: $10a^2b^4 + 15a^3b^2$

2. Factor: $3x^5y^{10} - 9x^7y^4 + 21x^2y^{12}$

3. Factor: $65y^9v^{18} + 20y^{30}v^{20} + 30y^{18}v^4$

Factor using Grouping

Factor: $3(x + 2) - x(x + 2)$

Example:

Factor $-2ys - 2yf + bs - bf$

Method 1 will not work because you can not pull out the GCF from all of the terms.
Since there are 4 terms, we can group 2 in common and another 2 in common.
$2ys - 2yf$ and $bs - bf$

Now, we can pull out the GCF from each group:
$2ys - 2yf$ and $bs - bf$

Pull out what they both have in common and see what remains.
$2y(s - f) + b(s - f)$
$(s - f)(2y + b)$, this is the factored form.
You can check by multiplying.

Examples:

1. Factor: $xy + 3y - 6x - 18$

2. Factor: $10k - 10m - km + m^2$

3. Factor: $x^2 - ax - x + a$

4. Factor: $5bx + 5bz + 25x + 25z$

What is factoring?
_____
_____

5.1: FACTORING BY GCF AND GROUPING PRACTICE PROBLEMS

1. Find the GCF of:
   54, 72, 90

2. Find the GCF of:
   x^{10}, x^{15}

3. Find the GCF of:
   $30x^6, 45x^9, 60x^{12}$

4. Find the GCF of:
   $36a^5b, 60a^7b^4, 120a^9b^6$

5. Factor:
   $12x^5 - 24x^3$

6. Factor:
   $24a^4b^6 + 40a^{10}b^5$

7. Factor:
   $18a^3b^7 + 36a^4b^2 - 9ab$

8. Factor:
   $40x^{10}y^{15}z^7 - 20x^7y^{30} - 60x^{22}y^{10}$

9. Factor: $x(x+3)+4(x+3)$

10. Factor: $x^2(2x-7)-3x(2x-7)+2(2x-7)$

11. Factor: $ax+ay+bx+by$

12. Factor: $x^2+x-2x-2$

13. Factor: $xy-xz-ay+az$

14. Factor: $15x^2-10xy+6x-4y$

15. Factor: $2a^2-3ab-2a+3b$

16. Factor: $20x^2+10xy-16x-8y$

FACTORING/RATIONAL EXPRESSION 185

5.2:
FACTORING TRINOMIALS IN THE FORM OF $x^2 + bx + c$

How do you factor trinomials in the form of $x^2 + bx + c$?

We will start by reviewing multiplication:

Simplify: $(x + 3)(x + 4)$

SOLUTION:

$(x + 3)(x + 4)$

F O I L

$x \cdot x + 4x + 3x + 3 \cdot 4$

$x^2 + 4x + 3x + 12$

F O + I L

$x^2 + 7x + 12$

To get the first term, it is the **multiplication** of the F part of FOIL

To get the inside term, it is the **addition** of the O + I of FOIL

To get the outside term, it is the **multiplication** of L of FOIL.

Simplify: $(x - 2)(x + 5)$

Example:

Factor- $x^2 + 7x + 12$

In this case pulling out the GCF method or the Grouping method will not work because you can not pull out the GCF from all of the terms and there are not an even number of terms to group.

Our goal is to factor $x^2 + 7x + 12$ into two factors that are being multiplied: ( )( )

To get the x^2, the factors must be $x \cdot x$ in the front (F of FOIL).

To factor this polynomial where the leading coefficient is 1:

We need two numbers that will multiply to the last term of: + 12 (L of FOIL) and add to the middle coefficient of: + 7 (O + I of FOIL)

We look at the factors of +12 and find: 1·12, 2·6, 3·4
−1· −12,   −2· −6,   −3· −4

Which pair will multiply to +12 and add to +7?

The numbers are: +3 and +4. (3)(4) = 12   3 + 4 = 7

The factored form is: $(x + 3)(x + 4)$

You can check by multiplying.

$(x + 3)(x + 4)$
$x \cdot x + 4x + 3x + 3 \cdot 4$
$x^2 + 4x + 3x + 12$
$x^2 + 7x + 12$

Know Your Signs

If you need two numbers which product is a positive number then the signs of the numbers are:

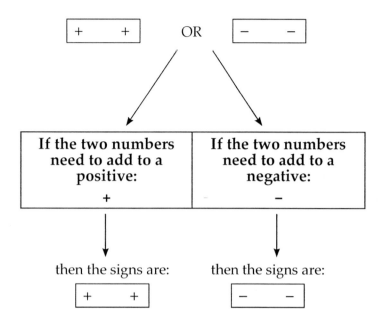

If you need two numbers which product is a negative number then the signs of the numbers are:

Examples:

1. Factor: $x^2 + 7x + 10$

2. Factor: $x^2 - 9x + 18$

3. Factor: $x^2 - 2x + 24$

4. Factor: $x^2 + 29x - 30$

5. Factor: $-x^2 + 4x + 45$

6. Factor: $2x^2 - 12x + 16$

How do we factor trinomials in the form $x^2 + bx + c$?

_____

_____

5.2: FACTORING TRINOMIALS OF THE FORM $x^2 + bx + c$ PRACTICE EXAMPLES

1. Factor: $x^2 + 4x + 3$

2. Factor: $x^2 + 9x + 14$

3. Factor: $x^2 - 6x + 9$

4. Factor: $x^2 - 8x + 12$

5. Factor: $x^2 - 5x - 24$

6. Factor: $x^2 - x - 20$

7. Factor: $x^2 + 3x - 18$

8. Factor: $x^2 + 6x - 27$

9. Factor: $x^2 - 2x - 1$

10. Factor: $35 + x^2 - 12x$

11. Factor: $x^2 + 23x - 50$

12. Factor: $x^2 + 7x - 12$

13. Factor: $x^2 + 27x + 72$

14. Factor: $x^2 - 9x - 36$

15. Factor: $-x^2 - 7x + 8$

16. Factor: $3x^2 - 3x - 18$

5.3:
FACTORING TRINOMIALS IN THE FORM OF $ax^2 + bx + c$

How do we factor trinomials in the form of $ax^2 + bx + c$?

Example:

Factor: $12x^2 - 23x + 5$
Method 1 or Method 2 or Method 3 will not work because: * You can not pull out of all of the terms (Method 1 Pull out the GCF), * There is not an even number of terms to group (Method 2 Grouping) and, * Since the leading coefficient is not 1, (Method 3 Trinomial a = 1). You still need two numbers that multiply to the last term, but they will NOT add to the middle term coefficient because the leading coefficient is not 1. There are several ways to factor this polynomial. Below you will find three different methods. <u>Trial and Error Method, Change to grouping Method, Square Root Method</u> **NOTE:** People prefer different methods. You can choose your preference or learn all of them.

First Method: Trial and Error
We know that when we factor $12x^2 - 23x + 5$, the factored form will be: ( )( )
<u>Step 1</u>: We will first list the possible factors: We must get the leading term of $12x^2$ The possibilities are: $(1x)(12x)$ $(2x)(6x)$ $(3x)(4x)$
<u>Step 2</u>: We know that it must multiply to the last term of + 5: The possibilities with signs are: $(+1)(12x)$ $(-1)(-5)$ Since the middle term is negative, and it must add to the middle term, then it must be: $(-1)(-5)$
<u>Step 3</u>: Then try different combinations and simplify each one to try to get: $12x^2 - 23x + 5$ Ex. $(2x - 5)(6x - 1)$, $(x - 5)(12x - 1)$, $(3x - 1)(4x - 5)$, $(3x - 5)(4x - 1)$ $(3x - 5)(4x - 1)$, this is the factored form. You can check by multiplying.

Second Method: Change to grouping

Example: Factor- $12x^2 - 23x + 5$

In this method we want the trinomial to become a polynomial with four terms so we can perform grouping. There is a specific way to change the trinomial into a polynomial with four terms.

Step 1: Find the middle factors.

Multiply the leading coefficient by the last constant term.

$12 \cdot 5 = +60$, also we want to find factors that multiply to get this number (including the sign).

The factors must also add to the middle term coefficient, in this case -23.

The factors that multiply to +60 and add to -23 are: -20 and -3.

Step 2: Rewrite the polynomial

We found the two coefficients to replace the middle term to form a polynomial with 4 terms.

The trinomial $12x^2 - 23x + 5$ will now become:

$$12x^2 - 3x - 20x + 5$$

It does not matter which number goes first, but since we are going to group, you want to place factors next to terms they have something in common with.

For example, place the "-3x" next to the "$12x^2$" because they have a "3x" in common and place the "-20x" next to the "5" because they have a "5" in common.

Step 3: Use Grouping
$12x^2 - 3x - 20x + 5$
$12x^2 - 3x \quad -20x + 5$
$3x(4x - 1) \quad -5(4x - 1)$
$(4x - 1)(3x - 5)$

$(4x - 1)(3x - 5)$, this is the factored form. You can check by multiplying.

FACTORING/RATIONAL EXPRESSION

Third Method: Square Root
Example: Factor- $12x^2 - 23x + 5$
Step 1: Make sure the trinomial is in the form: $ax^2 + bx + c$. $12x^2 - 23x + 5$
Step 2: Multiply the trinomial by the leading coefficient of (a = 12). $12(12x^2 - 23x + 5)$ $144x^2 - 276x + 60$
Step 3: Take the square root of the new leading term. This will become the first terms in the factored form. **Note:** Square roots will be treated in Ch. 6. $\sqrt{144x^2}$ $(12x\ )(12x\ )$
Step 4: To determine the other terms in the factored form you need to find two numbers that multiple to the **NEW** last term and add to the **ORIGINAL** middle number. In this case you need two numbers that multiply to +60 and add to -23. The numbers including the signs are (-20) and (-3). These numbers will be the second part in the parentheses. The order will not matter because the first part is 12x in both parentheses. $(12x - 20)(12x - 3)$
Step 5: The last step is to simplify so the original number that we multiplied by is divided out. We multiplied the trinomial by 12 so we need to divide the factored form by the factors of 12. The factors of 12 that divide into each part of the factored form evenly are 4 and 3. **Note:** The other factors of 12 such as 2 and 6 and 1 and 12 will not divide out evenly. $\dfrac{(12x-20)}{4} \qquad \dfrac{(12x-3)}{3}$ $\dfrac{12x}{4} - \dfrac{20}{4} \qquad \dfrac{12x}{3} - \dfrac{3}{3}$ $3x - 5 \qquad\qquad 4x - 1$ $\qquad(3x-5)(4x-1)$
$(3x - 5)(4x - 1)$. This is the factored form. You can check by multiplying.

Examples:

1. Factor: $2x^2 + 5x + 3$

2. Factor: $9x^2 - 9x + 2$

3. Factor: $10x^2 + 13x - 3$

4. Factor: $3x^2 - 7x - 6$

5. Factor: $-8x^3 + 2x^2 + 3x$

6. Factor: $12x^5 - 17x^4 + 6x^3$

How do we factor trinomials in the form $ax^2 + bx + c$?

_____

_____

5.3: FACTORING TRINOMIALS OF THE FORM $ax^2 + bx + c$ PRACTICE EXAMPLES

1. Factor: $2x^2 + 7x + 3$

2. Factor: $6x^2 + 7x + 2$

3. Factor: $8x^2 + 10x - 3$

4. Factor: $6x^2 - x - 12$

5. Factor: $4x^2 + 4x - 15$

6. Factor: $5x^2 - 17x + 6$

7. Factor: $25x^2 + 21x - 4$

8. Factor: $12x^2 - 25x + 12$

9. Factor: $20x^2 - 31x - 7$

10. Factor: $2x^2 + 3x + 9$

11. Factor: $2x^3 + 4x^2 - 30x$

12. Factor: $-3x^2 + 9x + 54$

5.4: FACTORING THE DIFFERENCE OF TWO SQUARES AND PERFECT SQUARE TRINOMIALS

How do we factor the difference of two squares?

Binomial (+/-) Special Case (Difference of Two Squares)

Example:

Factor: $4x^2 - 9$
Methods 1,2,3 and 4 will not work because you can not pull out of all of the terms, there are not an even number of terms to group, and it is not a trinomial.
This is a Special Case because: 1. There are only 2 terms 2. There is a negative sign in the middle 3. You can take the square root of both terms If all the criteria above represented, then you have a **difference of squares**.
To factor this polynomial, you take the square root of both terms: $2x$ and 3 Since there is no middle term, it must have canceled each other out. Therefore, you make one factor positive and another factor negative.
$(2x + 3)(2x - 3)$, also this is the factored form.
You can check by multiplying. $(2x + 3)(2x - 3)$ $4x^2 - 6x + 6x - 9$ $4x^2 - 9$

Examples:

1. Factor: $100 - x^2$

2. Factor: $49x^2 - 36$

3. Factor: $16x^2 - 25$

4. Factor: $8x^2 - 50$

5. Factor: $36x^2 - 81y^2$

6. Factor: $a^2 - b^2$

How do we factor the difference of two squares?

_____

_____

Factoring Perfect Square Trinomial

Example:

Factor: $x^2 + 10x + 25$

$x^2 + 10x + 25$ is a perfect square trinomial because:
     The first term: x^2 is the square of x: $x \cdot x = x^2$
     The last term: 25 is the square of 5: $5^2 = 25$
     The second term: 10x is twice the product of x and 5: $2(5)(x) = 10x$

A trinomial is a perfect square trinomial if:
     $x^2 + 10x + 25$
   $\sqrt{x^2} + 10x + \sqrt{25}$
     x   5
        $+2(5x)$
     x   +   5
      $(x+5)^2$

The factored form of: $x^2 + 10x + 25$ is: $(x+5)(x+5) = (x+5)^2$

Example:

Factor: $9x^2 - 42x + 49$

$9x^2 - 42x + 49$ is a perfect square trinomial because:
 	
   The first term: $9x^2$ is the square of 3x: $3x \cdot 3x = 9x^2$
   The last term: 49 is the square of 7: $7^2 = 49$
   The second term: -42x is - twice the product of 3x and 7: $-2(3x)(7) = -42x$

A trinomial is a perfect square trinomial if:
                  $9x^2 - 42x + 49$
              $\sqrt{9x^2} - 42x + \sqrt{49}$
                3x   7
                      $-2(3x)(7)$
                3x   −   7
 	
                  $(3x-7)^2$

The factored form of: $9x^2 - 42x + 49$ is: $(3x-7)(3x-7) = (3x-7)^2$

Examples:

1. Factor: $x^2 + 6x + 9$

2. Factor: $x^2 - 12x - 36$

3. Factor: $25x^2 - 30x + 9$

4. Factor: $9x^2 - 30xy + 25y^2$

5.4: FACTORING THE DIFFERENCE OF TWO SQUARES AND FACTOR PERFECT SQUARE TRINOMIALS - PRACTICE PROBLEMS

Factor the following:

1. $x^2 - 16$
2. $25 - x^2$
3. $9x^2 - 49$

4. $x^2 + 4$
5. $-x^2 + 100$
6. $2x^2 - 8$

7. $4x^2 - 36$
8. $c^2 - d^2$
9. $121x^2 - 81y^4$

10. $x^2 + 18x + 81$

11. $x^2 + 20x + 100$

12. $4x^2 + 28x + 49$

13. $x^2 - 10x + 25$

14. $x^2 - 8x - 16$

15. $9x^2 - 12x + 4$

16. $25x^2 + 60x + 36$

17. $49x^2 - 56x + 16$

18. $9x^2 - 66xy + 121y^2$

5.5: FACTORING USING MULTIPLE METHODS

If sometimes we need to factor using multiple methods, what process should we follow?

Steps for Factoring a Polynomial

Check each method in this order:

1. Pull out the GCF

    Ex. $25x^3y^3 + 15x^5y^6 - 5x^2y^2$

    $5x^2y^2(5xy + 3x^3y^4 - 1)$

2. Grouping (Hint: 4 terms or terms that can be paired)

    Ex. $2ys - 2yf + bs - bf$

    $(s - f)(2y + b)$

3. Trinomial where leading coefficient is 1

    Ex. $x^2 + 7x + 12$

    $(x + 3)(x + 4)$

4. Trinomial where the leading coefficient is not 1

    Ex. $12x^2 - 23x + 5$

    $(3x - 5)(4x - 1)$

5. Difference of Squares

    Ex. $4x^2 - 9$

    $(2x + 3)(2x - 3)$

6. Perfect Square Trinomials

    Ex. $4x^2 - 12x + 9$

    $(2x - 3)^2$

If sometimes we need to factor using multiple methods, what process should we follow?

_____

_____

Examples:

1. Factor: $8x^2 - 98$

2. Factor: $-4x^5y^2 - 12x^4y^3 - 9x^3y^4$

3. Factor: $-4x^2 + 32x + 4x^3$

4. Factor: $-24x^3y + 42x^2y^2 + 12y^3x$

5.5: FACTORING USING MULTIPLE METHODS. PRACTICE PROBLEMS

1. Factor: $7x^2$

2. Factor: $2x^3 - 16x^2 + 24x$

3. Factor: $x^2(a+b) - 25(a+b)$

4. Factor: $4x^3 + 12x^2 - 9x - 27$

5. Factor: $-6x - 5x^2 + 6x^3$

6. Factor: $a^2bc^2 - 10a^2bc + 24a^2b$

5.6: SOLVE QUADRATIC EQUATIONS BY FACTORING

How do we solve quadratic equations by factoring?

Example:

Solve: $2x^2 + 5x - 3 = 0$ by Factoring
In this method, MAKE SURE that the quadratic polynomial is set equal to 0.

Step 1: Factor the quadratic equation
$2x^2 + 5x - 3 = 0$ **Note**: We set each factor equal to zero $(2x - 1)(x + 3) = 0$ because multiplying by zero results in 0.

Step 2: Set each factor equal to 0
$(2x - 1)(x + 3) = 0$
$2x - 1 = 0 \quad x + 3 = 0$

Step 3: Solve each equation
$2x - 1 = 0$ $+1 \ \ +1$ $2x = 1 \qquad\qquad x + 3 = 0$ $\div 2 \ \ \ \div 2 \qquad\qquad\ \ -3 \ \ -3$ $\ \ x \ \ = \dfrac{1}{2} = 0.5 \qquad x = -3$

Step 4: Check your answers
The answers are <u>x = 0.5</u> and <u>x = -3</u>
Substitute each answer individually into the original equation:
$\quad 2x^2 + 5x - 3 = 0 \qquad\qquad 2x^2 + 5x - 3 = 0$ $2(0.5)^2 + 5(0.5) - 3 = 0 \qquad\ \ 2(-3)2 + 5(-3)-3 = 0$ $\qquad\qquad\qquad 0 = 0 \qquad\qquad\qquad\qquad\qquad 0 = 0$

Example:

Solve the quadratic equation: $x^2 - 4x - 12 = 0$
Step 1: Set the quadratic equation equal to zero (if needed)
Step 2: Factor the quadratic equation $(x - 6)(x + 2) = 0$
Step 3: Set each factor equal to zero $x - 6 = 0 \qquad x + 2 = 0$
Step 4: Solve each factor $\begin{array}{ll} x - 6 = 0 & x + 2 = 0 \\ +6 \quad +6 & -2 \quad -2 \\ x = 6 & x = -2 \end{array}$
Step 5: Check the solutions $\begin{array}{ll} x^2 - 4x - 12 = 0 & x^2 - 4x - 12 = 0 \\ (6)^2 - 4(6) - 12 = 0 & (-2)^2 - 4(-2) - 12 = 0 \\ 36 - 24 - 12 = 0 & 4 + 8 - 12 = 0 \\ \qquad\quad 0 = 0 & \qquad\quad 0 = 0 \end{array}$

Example:

Solve: $6x^2 - 7x + 2 = 0$ by factoring

FACTORING/RATIONAL EXPRESSION

Example:

Solve the quadratic equation: $3x(x + 1) = 2x + 2$.

Step 1: Set the quadratic equation equal to zero.

Step 2: Factor the quadratic equation.

Step 3: Set each factor equal to zero.

Step 4: Solve each factor.

Step 5: Check the solutions.

How do we solve quadratic equations by factoring?

_____

_____

5.6: SOLVE QUADRATIC EQUATIONS BY FACTORING - PRACTICE PROBLEMS

Solve:

1. $(x + 2)(x - 3) = 0$

2. $x(x + 6)(2x - 1) = 0$

3. $x^2 - 3x = 0$

4. $5x^2 + 10x = 0$

5. $4x(x + 7) - 3(x + 7) = 0$

6. $8x^2 + 12x + 10x + 15 = 0$

7. $x^2 - 11x + 30 = 0$

8. $x^2 - 100 = 0$

9. $12x^2 - 17x - 5 = 0$

10. $x^2 - 4x - 30 = 2$

11. $-10 + 6x^2 = 11x$

12. $14x^3 + 19x^2 + 5x = 8x$

5.7: FACTORING APPLICATIONS

Examples where factoring is used:

Review Example:

Find the area of a rectangle, given that the length of the rectangle is 5 more than twice the width.

SOLUTION:

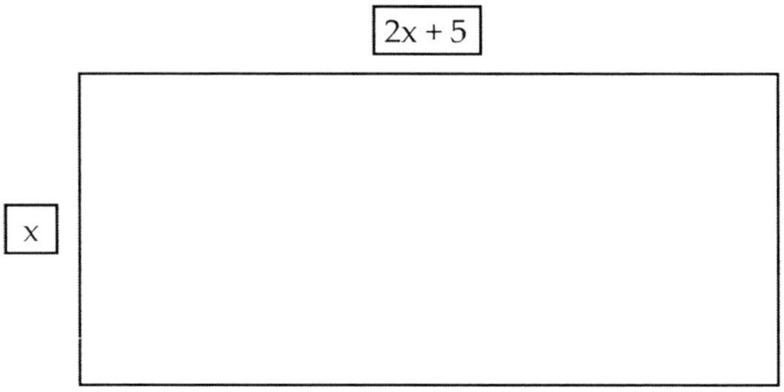

To find the area of a rectangle use the formula for the area of a rectangle:

Area = Length times Width or $A = L \cdot W$

In this case, the width is "x" and the length is "2x + 5" so the area is:

$$A = L \cdot W$$
$$A = x(2x + 5)$$
$$A = 2x^2 + 5x$$

Factoring Application 1: Factoring can be used to find unknowns.

Example:

Find the length and width of a rectangle whose area is: $2x^2 + 5x + 2 \, ft^2$

(In this case, the length is the longer side and the width is the shorter side.)

Once the width and length algebraic expressions are found, find the value of the width and length when x = 5 ft. Then verify the area by substituting x = 5 ft into the area expression.

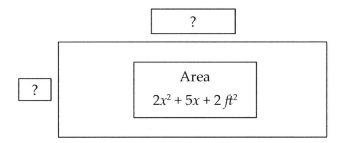

Solution:

To find the solution factor the area to find the length and width:

$$2x^2 + 5x + 2$$
$$(x + 2)(2x + 1)$$

The width is "x + 2" feet and the length "2x + 1" feet.

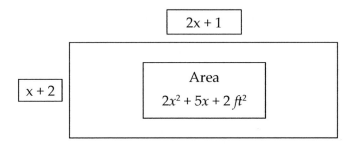

If x = 5 ft, then:

Width	Length	Area	Area
			$2x^2 + 5x + 2 \, ft^2$
x + 2	2x + 1		$2(5)2 + 5(5) + 2$
(5) + 2	2(5) + 1		50 + 25 + 2
7ft	11ft	(7 ft)(11 ft) = 77 ft²	77 ft²

Examples:

1. Find the length and width of a rectangular garden whose area is:

   $x^2 - 25 \; ft^2$

   (In this case the length is the longer side and the width is the shorter side.)

   Once the algebraic expressions for width and length are found, find the numerical value of the width and length when x = 10 ft. Then verify the area by substituting x = 10 ft into the algebraic expression for area.

   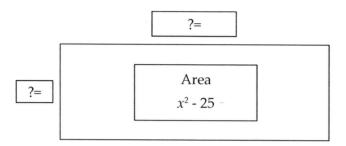

2. Find the length and width of a square play area whose area is: $4x^2 + 12x + 9 \; ft^2$.

   Once the algebraic expressions for width and length are found, find the numerical value of the width and length when x = 6 ft. Then verify the area by substituting x = 6 ft into the algebraic expression for area.

   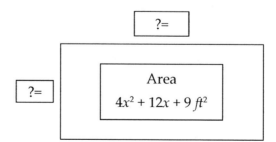

Factoring Application 2:

Example:

The equation of motion of a person who dives off a 48 ft. cliff into a river with an initial velocity of 32 ft/sec is:

$h = -16t^2 + 32t + 48$, where "t" is time in seconds and "h" is height in feet.

When will the diver hit the water (when height = 0)?

Solution:

To find the solution:

1. Substitute 0 in for the height (h)
   $h = -16t^2 + 32t + 48$
   $0 = -16t^2 + 32t + 48$
   $-16t^2 + 32t + 48 = 0$

2. Factor the trinomial by first factoring out the -16:
   $-16t^2 + 32t + 48 = 0$
   $-16(t^2 - 2t - 3) = 0$

3. Continue by factoring the trinomial:
   $-16(t^2 - 2t - 3) = 0$
   $-16(t - 3)(t + 1) = 0$

4. Set each term with a variable equal to the right side of 0 and solve for t:
   $t - 3 = 0$ $t + 1 = 0$
   $+3 +3$ $-1 -1$
   $t = 3$ $t = -1$

5. The solutions are 3 seconds and -1 seconds. Since -1 seconds is not realistic, we do not include that solution. The diver will hit the water in 3 seconds.

Examples:

1. The equation of motion of a person who dives off a 64 ft. cliff into a river with an initial velocity of 0 ft/sec is:

   $h = -16t^2 + 64$, where "t" is time in seconds and "h" is height in feet.

   When will the diver hit the water (when height = 0)?

2. The equation of motion of the path of a ball kicked from the ground in the air at a speed of 96 ft/sec is: $h = -16t^2 + 96t$, where "t" is time (seconds) and "h" is height in feet.

   When will the ball hit the ground (when height = 0)?

What are some examples where factoring is used?

_____

_____

5.7: FACTORING APPLICATIONS. PRACTICE PROBLEMS

Solve:

1. Find the length and width of a rectangular room whose area is: $16x^2 - 25 \text{ ft}^2$.
   Draw a picture of the situation.

   (In this case the length is the longer side and the width is the shorter side.)
   Once the width and length are found, find the value of the width and length when $x = 10$ ft.
   Then verify the area by substituting $x = 10$ ft into the area.

2. Find the length and width of a square room whose area is: $4x^2 + 20x + 25 \text{ ft}^2$.
   Draw a picture of the situation.

   (If the length and width are different, then the length is the longer side.)
   Once the width and length are found, find the value of the width and length when $x = 4$ ft.
   Then verify the area by substituting $x = 4$ ft into the area.

3. The equation of motion of a person who dives off a 96 ft. cliff into a river with an initial velocity of 16 ft/sec is:
$h = -16t^2 + 16t + 96$, where "t" is time in seconds and "h" is height in feet.
When will the diver hit the water (when height = 0)?

4. The equation of motion of the path of a ball kicked in the air at a speed of 16 ft/sec is:
$h = -16t^2 + 16t$, where "t" is time (seconds) and "h" is height in feet.
When will the ball hit the ground (when height = 0)?

5.8: SIMPLIFY RATIONAL EXPRESSIONS

How do we simplify rational expressions?

Example:

Simplify- $\dfrac{10}{35}$

SOLUTION:

<u>Step 1</u>: Factor the numerator and denominator

$$\dfrac{10}{35} = \dfrac{2 \cdot 5}{7 \cdot 5}$$

<u>Step 2</u>: Simplify

$$\dfrac{2 \cdot \cancel{5}}{7 \cdot \cancel{5}}$$

<u>Step 3</u>: Final Answer

$$\dfrac{2}{7}$$

Examples:

1. Simplify- $\dfrac{45}{63}$

2. Simplify- $\dfrac{60}{140}$

3. Simplify- $\dfrac{25x^4y}{10x^2y^5}$

4. Simplify- $\dfrac{14a^7b^4}{20a^3b^5}$

Example:

Simplify- $\dfrac{3x^2 - 8x - 3}{x^2 - 9}$

SOLUTION:

Step 1: Factor the numerator and denominator

$$\dfrac{3x^2 - 8x - 3}{x^2 - 9} = \dfrac{(3x+1)(x-3)}{(x+3)(x-3)}$$

Step 2: Simplify

$$\dfrac{(3x+1)(x-3)}{(x+3)(x-3)}$$

Step 3: Final Answer

$$\dfrac{3x+1}{x+3}$$

Examples:

1. Simplify- $\dfrac{2x - 14}{x^2 - 49}$

2. Simplify- $\dfrac{x^2 + 3x + 2}{x^2 + x - 2}$

3. Simplify- $\dfrac{x^2-4}{x^2-2x-8}$

4. Simplify- $\dfrac{x^2+5x+6}{x^2-x-6}$

5. Simplify- $\dfrac{6x-30}{5-x}$

6. Simplify- $\dfrac{6x^2-13x+6}{x^2+4x+4}$

How do we simplify rational expressions?

_____
_____

5.8: SIMPLIFY RATIONAL EXPRESSIONS - PRACTICE PROBLEMS

Simplify:

1. $\dfrac{7}{77}$

2. $\dfrac{50}{75}$

3. $\dfrac{21x^3y^2}{6xy^4}$

4. $\dfrac{36x^5y^3}{72x^4y^3}$

5. $\dfrac{x^2 - 3x}{x - 3}$

6. $\dfrac{5x}{25x^2 - 30x}$

7. $\dfrac{x^2 + 2x}{x^2 + 3x + 2}$

8. $\dfrac{x^2 + 3x + xy + 3y}{x^2 + 5x + 6}$

9. $\dfrac{x+3}{x^2+x-6}$

10. $\dfrac{x^2-2x-15}{x^2-8x+15}$

11. $\dfrac{x^2-16}{x^2-8x+16}$

12. $\dfrac{2x^2+11x+5}{2x^2-50}$

13. $\dfrac{3x^2+13x+12}{x^2-4x-21}$

14. $\dfrac{6x^2-7x-3}{8x^2-6x-9}$

5.9: MULTIPLY AND DIVIDE RATIONAL EXPRESSIONS

How do we multiply and divide rational expressions?

Review Examples:

Simplify: $\dfrac{10}{35}$

Solution:

Step 1: Factor the numerator and denominator

$$\dfrac{10}{35} = \dfrac{2 \cdot 5}{7 \cdot 5}$$

Step 2: Simplify

$$\dfrac{2 \cdot \cancel{5}}{7 \cdot \cancel{5}} = \dfrac{2}{7}$$

Simplify: $\dfrac{x^5}{x^3}$

Solution:

Step 1: Factor the numerator and denominator

$$\dfrac{x^5}{x^3} = \dfrac{x \cdot x \cdot x \cdot x \cdot x}{x \cdot x \cdot x}$$

Step 2: Simplify

$$\dfrac{x^5}{x^3} = \dfrac{\cancel{x} \cdot \cancel{x} \cdot \cancel{x} \cdot x \cdot x}{\cancel{x} \cdot \cancel{x} \cdot \cancel{x}} = x^2$$

Simplify: $\dfrac{x^3 - 8x^2 - 2x}{x^2 - x}$

Solution:

Step 1: Factor the numerator and denominator

$$\dfrac{x^3 - 8x^2 - 2x}{x^2 - x} = \dfrac{x(x^2 - 8x - 2)}{x(x-1)}$$

Step 2: Simplify

$$\dfrac{\cancel{x}(x^2 - 8x - 2)}{\cancel{x}(x-1)} = \dfrac{x^2 - 8x - 2}{x - 1}$$

Simplify: $\dfrac{3x^2 - 8x - 3}{x^2 - 9}$

Solution:

Step 1: Factor the numerator and denominator

$$\dfrac{3x^2 - 8x - 3}{x^2 - 9} = \dfrac{(3x+1)(x-3)}{(x+3)(x-3)}$$

Step 2: Simplify

$$\dfrac{(3x+1)(x-3)}{(x+3)(x-3)} = \dfrac{3x+1}{x+3}$$

FACTORING/RATIONAL EXPRESSION

Now we are going to include multiplication:

Multiply: $\dfrac{5}{14} \cdot \dfrac{21}{10}$

Solution:

<u>Step 1</u>: Factor the numerators and denominators

$$\dfrac{5}{14} \cdot \dfrac{21}{10} = \dfrac{5}{2 \cdot 7} \cdot \dfrac{3 \cdot 7}{2 \cdot 5}$$

<u>Step 2</u>: Simplify then multiply

$$\dfrac{\cancel{5}}{2 \cdot \cancel{7}} \cdot \dfrac{3 \cdot \cancel{7}}{2 \cdot \cancel{5}} = \dfrac{3}{4}$$

Multiply: $\dfrac{x^5}{y^2} \cdot \dfrac{y}{x^3}$

Solution:

<u>Step 1</u>: Factor the numerators and denominators

$$\dfrac{x^5}{y^2} \cdot \dfrac{y}{x^3} = \dfrac{x \cdot x \cdot x \cdot x \cdot x}{y \cdot y} \cdot \dfrac{y}{x \cdot x \cdot x}$$

<u>Step 2</u>: Simplify then multiply

$$\dfrac{\cancel{x} \cdot \cancel{x} \cdot \cancel{x} \cdot x \cdot x}{y \cdot \cancel{y}} \cdot \dfrac{\cancel{y}}{\cancel{x} \cdot \cancel{x} \cdot \cancel{x}} = \dfrac{x^2}{y}$$

Multiply: $\dfrac{x^3 - 2x^2}{5x+5} \cdot \dfrac{5}{3x-6}$

Solution:

<u>Step 1</u>: Factor the numerators and denominators

$$\dfrac{x^3 - 2x^2}{5x+5} \cdot \dfrac{5}{3x-6} = \dfrac{x^2(x-2)}{5(x+1)} \cdot \dfrac{5}{3(x-2)}$$

<u>Step 2</u>: Simplify then multiply

$$\dfrac{x^2\cancel{(x-2)}}{\cancel{5}(x+1)} \cdot \dfrac{\cancel{5}}{3\cancel{(x-2)}} = \dfrac{x^2}{3(x+1)} = \dfrac{x^2}{3x+3}$$

Multiply: $\dfrac{x^2+5x+6}{x^2-16} \cdot \dfrac{2x^2+9x+4}{x^2+2x}$

Solution:

<u>Step 1</u>: Factor the numerators and denominators

$$\dfrac{x^2+5x+6}{x^2-16} \cdot \dfrac{2x^2+9x+4}{x^2+2x} =$$

$$\dfrac{(x+2)(x+3)}{(x-4)(x+4)} \cdot \dfrac{(2x+1)(x+4)}{x(x+2)}$$

<u>Step 2</u>: Simplify then multiply

$$\dfrac{\cancel{(x+2)}(x+3)}{(x-4)\cancel{(x+4)}} \cdot \dfrac{(2x+1)\cancel{(x+4)}}{x\cancel{(x+2)}} = \dfrac{(x+3)(2x+1)}{x(x-4)}$$

$$= \dfrac{2x^2+7x+3}{x^2-4x}$$

Now we are going to include division:

Divide: $\dfrac{2}{15} \div \dfrac{6}{35}$

Solution:

Step 1: Convert to multiplication by taking the reciprocal of the second expression

$$\dfrac{2}{15} \div \dfrac{6}{35} = \dfrac{2}{15} \cdot \dfrac{35}{6}$$

Step 2: Factor the numerators and denominators

$$\dfrac{2}{15} \cdot \dfrac{35}{6} = \dfrac{2}{3 \cdot 5} \cdot \dfrac{5 \cdot 7}{2 \cdot 3}$$

Step 3: Simplify then multiply

$$\dfrac{\cancel{2}}{3 \cdot \cancel{5}} \cdot \dfrac{\cancel{5} \cdot 7}{\cancel{2} \cdot 3} = \dfrac{7}{9}$$

Divide: $\dfrac{x^2}{y^3 z} \div \dfrac{x}{y^5 z}$

Solution:

Step 1: Convert to multiplication by taking the reciprocal of the second expression

$$\dfrac{x^2}{y^3 z} \div \dfrac{x}{y^5 z} = \dfrac{x^2}{y^3 z} \cdot \dfrac{y^5 z}{x}$$

Step 2: Factor the numerators and denominators

$$\dfrac{x^2}{y^3 z} \cdot \dfrac{y^5 z}{x} = \dfrac{x \cdot x}{y \cdot y \cdot y \cdot z} \cdot \dfrac{y \cdot y \cdot y \cdot y \cdot y \cdot z}{x}$$

Step 3: Simplify then multiply

$$\dfrac{x \cdot \cancel{x}}{\cancel{y} \cdot \cancel{y} \cdot \cancel{y} \cdot \cancel{z}} \cdot \dfrac{\cancel{y} \cdot \cancel{y} \cdot \cancel{y} \cdot y \cdot y \cdot \cancel{z}}{\cancel{x}} = \dfrac{xy^2}{1} = xy^2$$

Divide: $\dfrac{x^2 + 3x}{5y^2} \div \dfrac{2x}{y^2 + 2y}$

Solution:

Step 1: Convert to multiplication by taking the reciprocal of the second expression

$$\dfrac{x^2 + 3x}{5y^2} \div \dfrac{2x}{y^2 + 2y} = \dfrac{x^2 + 3x}{5y^2} \cdot \dfrac{y^2 + 2y}{2x}$$

Step 2: Factor the numerators and denominators

$$\dfrac{x^2 + 3x}{5y^2} \cdot \dfrac{y^2 + 2y}{2x} = \dfrac{x(x+3)}{5 \cdot y \cdot y} \cdot \dfrac{y(y+2)}{2x}$$

Step 3: Simplify then multiply

$$\dfrac{x(x+3)}{5 \cdot y \cdot y} \cdot \dfrac{y(y+2)}{2x} = \dfrac{(x+3)(y+2)}{10y}$$

$$= \dfrac{xy + 2x + 3y + 6}{10y}$$

Divide: $\dfrac{x^2 + 3x + 2}{x^2 - 3x} \div \dfrac{2x^2 + 5x + 2}{x^2 - 9}$

Solution:

Step 1: Convert to multiplication by taking the reciprocal of the second expression

$$\dfrac{x^2 + 3x + 2}{x^2 - 3x} \div \dfrac{2x^2 + 5x + 2}{x^2 - 9} =$$
$$\dfrac{x^2 + 3x + 2}{x^2 - 3x} \cdot \dfrac{x^2 - 9}{2x^2 + 5x + 2}$$

Step 2: Factor the numerators and denominators

$$\dfrac{x^2 + 3x + 2}{x^2 - 3x} \cdot \dfrac{x^2 - 9}{2x^2 + 5x + 2}$$

$$\dfrac{(x+1)(x+2)}{x(x-3)} \cdot \dfrac{(x-3)(x+3)}{(2x+1)(x+2)}$$

Step 3: Simplify then multiply

$$\dfrac{(x+1)(x+2)}{x(x-3)} \cdot \dfrac{(x-3)(x+3)}{(2x+1)(x+2)} = \dfrac{(x+1)(x+3)}{x(2x+1)}$$

$$= \dfrac{x^2 + 4x + 3}{2x^2 + x}$$

Examples:

1. Multiply: $\dfrac{3}{28} \cdot \dfrac{14}{15}$

2. Multiply: $\dfrac{x^5}{y^3 z} \cdot \dfrac{yz^4}{x^2}$

3. Multiply: $\dfrac{2x^4 - 6x^2}{7x + 14} \cdot \dfrac{7}{2x}$

4. Multiply: $\dfrac{3x^4 + 5x^2}{6x^2 + 13x + 6} \cdot \dfrac{4x^2 - 9}{x^3}$

Examples:

1. Divide: $\dfrac{6}{25} \div \dfrac{18}{35}$

2. Divide: $\dfrac{5x}{y^3} \div \dfrac{15x}{y^2 z}$

3. Divide: $\dfrac{3y^2}{5y^2 + y} \div \dfrac{y^2 + 2y}{5y + 1}$

4. Divide: $\dfrac{x^2 - x - 42}{2x^2 - 13x - 7} \div \dfrac{x^2 + 6x}{x^2 - 49}$

How do we multiply and divide rational expressions?

_____

_____

5.9 MULTIPLY AND DIVIDE RATIONAL EXPRESSIONS PRACTICE EXAMPLES

1. Multiply: $\dfrac{20}{6} \cdot \dfrac{6}{4}$

2. Multiply: $\dfrac{a^3}{b^5 c^2} \cdot \dfrac{b^2}{ac^2}$

3. Multiply: $\dfrac{x^2 + x}{4x + 8} \cdot \dfrac{x + 2}{x}$

4. Multiply: $\dfrac{x^2 + 5x}{x^2 + 8x + 15} \cdot \dfrac{3x^2 + 13x + 12}{x^2 - 25}$

5. Divide: $\dfrac{11}{14} \div \dfrac{33}{28}$

6. Divide: $\dfrac{3a}{c^4} \div \dfrac{9a^2b}{c^5}$

7. Divide: $\dfrac{3x+2}{x} \div \dfrac{6x^2+4x}{x^3+2x^2}$

8. Divide: $\dfrac{x^2-6x+8}{2x^2+2x} \div \dfrac{x^2-16}{2x^2+5x+3}$

5.10: ADD AND SUBTRACT RATIONAL EXPRESSIONS WITH MONOMIAL DENOMINATORS

$$\frac{5}{7x} + \frac{2}{7x}$$

How do we add and subtract rational expressions with monomial denominators?

Review Examples:

Adding/Subtracting Like Denominators	Adding/Subtracting Unlike Denominators
Subtract: $\dfrac{5}{7} - \dfrac{2}{7}$	Add: $\dfrac{2}{3} + \dfrac{1}{4}$
SOLUTION:	**SOLUTION:**
Add or subtract the numerator while keeping the denominator the same, then simplify if needed $$\frac{5}{7} - \frac{2}{7} = \frac{5-2}{7} = \frac{3}{7}$$	**Step 1**: Find the common denominator $$\frac{2}{3} + \frac{1}{4}$$ $$\frac{}{12} + \frac{}{12}$$ **Step 2**: Convert to equivalent fractions $$\frac{2 \cdot 4}{3 \cdot 4} + \frac{1 \cdot 3}{4 \cdot 3}$$ $$\frac{8}{12} + \frac{3}{12}$$ **Step 3**: Add or subtract, then simplify if needed $$\frac{8}{12} + \frac{3}{12} = \frac{8+3}{12} = \frac{11}{12}$$

Adding/Subtracting Rational Expressions with like denominators

Since the denominators are the same, we add/subtract the numerators and write the sum or difference over the common denominator. Then simplify the expression, if needed.

Add: $\dfrac{x}{3}+\dfrac{2}{3}$ Solution: $\dfrac{x}{3}+\dfrac{2}{3}=\dfrac{x+2}{3}$	Subtract: $\dfrac{x}{5}-\dfrac{3}{5}$ Solution: $\dfrac{x}{5}-\dfrac{3}{5}=\dfrac{x-3}{5}$
Add: $\dfrac{2}{x}+\dfrac{3}{x}$ Solution: $\dfrac{2}{x}+\dfrac{3}{x}=\dfrac{2+3}{x}=\dfrac{5}{x}$	Subtract: $\dfrac{y}{7x}-\dfrac{3}{7x}$ Solution: $\dfrac{y}{7x}-\dfrac{3}{7x}=\dfrac{y-3}{7x}$
Add: $\dfrac{x+7}{2x^2}+\dfrac{3}{2x^2}$ Solution: $\dfrac{x+7}{2x^2}+\dfrac{3}{2x^2}=\dfrac{x+7+3}{2x^2}=\dfrac{x+10}{2x^2}$	Subtract: $\dfrac{5x+2}{6y}-\dfrac{3x+1}{6y}$ Solution: $\dfrac{5x+2}{6y}-\dfrac{3x+1}{6y}=\dfrac{5x+2-(3x+1)}{6y}=\dfrac{5x+2-3x-1}{6y}=\dfrac{2x+1}{6y}$

FACTORING/RATIONAL EXPRESSION

Adding/Subtracting Rational Expressions with unlike denominators

When the denominators are not the same, we follow these steps:

Step 1: Find the common denominator.

Step 2: Convert to equivalent fractions.

Step 3: Add or subtract, then simplify if needed.

Subtract: $\dfrac{2}{5} - \dfrac{3}{x}$

SOLUTION:

Step 1: Find the common denominator

$$\dfrac{2}{5} - \dfrac{3}{x}$$

$$\dfrac{}{5x} - \dfrac{}{5x}$$

Step 2: Convert to equivalent fractions

$$\dfrac{2 \cdot x}{5 \cdot x} - \dfrac{3 \cdot 5}{x \cdot 5}$$

$$\dfrac{2x}{5x} - \dfrac{15}{5x}$$

Step 3: Add or subtract, then simplify if needed

$$\dfrac{2x}{5x} - \dfrac{15}{5x} = \dfrac{2x - 15}{5x}$$

Add: $\dfrac{2}{3x} + \dfrac{1}{4}$

SOLUTION:

Step 1: Find the common denominator

$$\dfrac{2}{3x} + \dfrac{1}{4}$$

$$\dfrac{}{12x} + \dfrac{}{12x}$$

Step 2: Convert to equivalent fractions

$$\dfrac{2 \cdot 4}{3x \cdot 4} + \dfrac{1 \cdot 3x}{4 \cdot 3x}$$

$$\dfrac{8}{12x} + \dfrac{3x}{12x}$$

Step 3: Add or subtract, then simplify if needed

$$\dfrac{8}{12x} + \dfrac{3x}{12x} = \dfrac{8 + 3x}{12x} \text{ or } \dfrac{3x + 8}{12x}$$

CHAPTER 5

Subtract: $\dfrac{2y}{x} - \dfrac{5}{y}$

Solution:

Step 1: Find the common denominator

$$\dfrac{2y}{x} - \dfrac{5}{y}$$

$$\dfrac{}{xy} - \dfrac{}{xy}$$

Step 2: Convert to equivalent fractions

$$\dfrac{2y \cdot y}{x \cdot y} - \dfrac{5 \cdot x}{y \cdot x}$$

$$\dfrac{2y^2}{xy} - \dfrac{5x}{xy}$$

Step 3: Add or subtract, then simplify if needed

$$\dfrac{2y^2}{xy} - \dfrac{5x}{xy} = \dfrac{2y^2 - 5x}{xy} \; or \; \dfrac{-5x + 2y^2}{xy}$$

Add: $\dfrac{y}{2x^2} + \dfrac{3}{4xy}$

Solution:

Step 1: Find the common denominator

$$\dfrac{y}{2x^2} + \dfrac{3}{4xy}$$

$$\dfrac{}{4x^2y} + \dfrac{}{4x^2y}$$

Step 2: Convert to equivalent fractions

$$\dfrac{y \cdot 2y}{2x^2 \cdot 2y} + \dfrac{3 \cdot x}{4xy \cdot x}$$

$$\dfrac{2y^2}{4x^2y} + \dfrac{3x}{4x^2y}$$

Step 3: Add or subtract, then simplify if needed

$$\dfrac{2y^2}{4x^2y} + \dfrac{3x}{4x^2y} = \dfrac{2y^2 + 3x}{4x^2y} \; or \; \dfrac{3x + 2y^2}{4x^2y}$$

Examples:

1. Add: $\dfrac{xy}{9} + \dfrac{z}{9} =$

2. Subtract: $\dfrac{y}{x^2} - \dfrac{6}{x^2} =$

3. Add: $\dfrac{x+3}{5y} + \dfrac{x+4}{5y}$

4. Subtract: $\dfrac{5}{3x^2} - \dfrac{x+2}{3x^2}$

5. Add: $\dfrac{3}{4} + \dfrac{2}{x}$

6. Subtract: $\dfrac{5}{x} - \dfrac{7}{y}$

7. Add: $\dfrac{7}{6x^2y} + \dfrac{3}{2xy^2}$

8. Subtract: $\dfrac{3y}{x^5} - \dfrac{5}{2x^2y}$

How do we add and subtract rational expressions with monomial denominators?

_____

_____

5.10 ADD AND SUBTRACT RATIONAL EXPRESSIONS WITH MONOMIAL DENOMINATORS - PRACTICE PROBLEMS

1. Add: $\dfrac{x}{3} + \dfrac{y}{3}$

2. Subtract: $\dfrac{3x}{5y^2} - \dfrac{x}{5y^2}$

3. Add: $\dfrac{2x^2}{yz^3} + \dfrac{3x^2+7}{yz^3}$

4. Subtract: $\dfrac{2x+1}{6x} - \dfrac{4x+5}{6x}$

5. Add: $\dfrac{1}{5}+\dfrac{4}{y}$

6. Subtract: $\dfrac{2}{xy}-\dfrac{8}{y}$

7. Add: $\dfrac{3}{4x^3}+\dfrac{7}{3x^2y}$

8. Subtract: $\dfrac{y}{2x^2}-\dfrac{6}{5xy}$

MAT0028C DEVELOPMENTAL MATH II   Name: _____

TEST 3 (CHAPTER 5) REVIEW
Questions from Chapter 5

Factor:

1. $10x^6 - 6x^8$

2. $24v^4w^4 + 4vw^2$

3. $45x^6y^{55} + 10x^{18}y^{15} + 55x^{24}y^{35}$

4. $2x^2 + 8x + 3x + 12$

5. $7x^3 - 21x^2 + x - 3$

6. $15y^2 - 10yz + 6y - 4z$

7. $x^2 + 11x + 30$

8. $x^2 - 8x + 16$

9. $x^2 - x - 6$

10. $4c^2 + 12c + 9$

11. $9n^2 - 9n - 10$

12. $12x^2 + 19x + 5$

13. $16z^2 - 25$

14. $2b^2 - 2c^2$

Solve:

15. $x^2 - 8x = 0$

16. $x^2 - 4x - 5 = 0$

Solve:

17. $3x^2 + 2x - 8 = 0$

18. $4x^2 + 3x - 1 = 0$

19. Simplify: $\dfrac{x^2 - x - 56}{x^2 - 49}$

20. Simplify: $\dfrac{x^2 - 4x - 12}{2x^2 - 15x + 18}$

21. Simplify: $\dfrac{3x^2+7x+2}{x} \cdot \dfrac{x^2+5x}{x^2+x-2}$

22. Simplify: $\dfrac{x^2+4x+3}{x+2} \div \dfrac{2x^2+5x-3}{x^2+2x}$

23. Simplify: $\dfrac{5}{2x} + \dfrac{3y}{3x^2}$

24. Simplify: $\dfrac{3}{2x^2y^4} - \dfrac{7z}{x^3y}$

Review Questions

25. Two machines can complete 8 tasks every 3 days. Let t represent the number of tasks these machines can complete in a 30-day month. Write a proportion to show this example.

26. Simplify: $7x + 8(x-3)$

27. Convert to scientific notation: 0.000023

28. Solve: $2x - 10 < -8$

29. Simplify: $(a^2b^4)(a^3b^4)$

30. Simplify: $|8 + (-14)| + 9$

31. Solve for t: $x = -8z + 7t$

32. Find the y–intercept for: $-5x + 8y = -8$

33. Graph: $y = 3x + 6$

34. The length of a rectangular pool is 6 less than twice the width. The perimeter of the pool is 78 feet. Find the length and width of the pool. Label each distance correctly.

RADICALS

CHAPTER 6

- **6.1: AN INTRODUCTION TO SQUARE ROOTS**
- **6.2: SIMPLIFY SQUARE ROOTS**
- **6.3: ADDING AND SUBTRACTING RADICAL EXPRESSIONS**
- **6.4: MULTIPLYING AND DIVIDING RADICAL EXPRESSIONS**
- **6.5: RATIONALIZING THE DENOMINATOR (MONOMIALS ONLY)**
- **6.6: SOLVING RADICAL EQUATIONS**
- **6.7: HIGHER ORDER ROOTS**

6.1: AN INTRODUCTION TO SQUARE ROOTS

what are square roots and when are they used?

Square Root of x– the real number that when multiplied by itself produces x

$\sqrt{}$ – radical symbol

$\sqrt{36}$ – the 36 in this example is the radicand

Example:

Solve: $\sqrt{36}$

Solution:

$\sqrt{36} = 6$ because $6 \cdot 6 = 36$

List of perfect squares:

1   4   9   16   25   36   49   64   81   100   121   144

Examples:

1. $\sqrt{4}$

2. $\sqrt{16}$

3. $\sqrt{1}$

4. $\sqrt{25}$

5. $\sqrt{121}$

6. $\sqrt{169}$

Estimate Square Roots

Example:

Estimate the $\sqrt{45}$

SOLUTION:

The $\sqrt{45}$ is between the $\sqrt{36}$ which is 6 and $\sqrt{49}$ which is 7.

So the $\sqrt{45}$ is approximately 6.7.

Estimate the following square roots.

1. $\sqrt{6}$

2. $\sqrt{115}$

3. $\sqrt{90}$

4. $\sqrt{55}$

Further Examples

1. The square root of a perfect square is a rational number.
   $\sqrt{100} = 10$

2. The square root of a positive non perfect square is an irrational number.
   $\sqrt{7} \approx 2.645751311....$

3. The square root of a negative number is a non real number called imaginary number.
   $\sqrt{-4}$
   $\sqrt{-4} = 2i$

Square Roots of Variable Expressions

$\sqrt{9} = 3$ because $\sqrt{9} = \sqrt{3 \cdot 3} = 3$

Since we have pair, the square root of 9 is 3.

The same is true for variables.

$\sqrt{x^2} = x$ because $\sqrt{x^2} = \sqrt{x \cdot x} = x$

$\sqrt{x^4} = x^2$ because $\sqrt{x^4} = \sqrt{(x \cdot x) \cdot (x \cdot x)} = x^2$

Find the pattern:

1. $\sqrt{x^6} =$ 　　　　　　2. $\sqrt{x^8} =$ 　　　　　　3. $\sqrt{x^0} =$

What is the pattern? _____

_____

Examples:

1. $\sqrt{x^{20}}$ 　　　　　　2. $\sqrt{25x^{14}}$

3. $\sqrt{49x^{18}y^6}$ 　　　　　　4. $\sqrt{36x^8}$

RADICALS 249

Pythagorean Theorem- In a right triangle, the sum of the squares of the lengths of the two shorter sides (legs) is equal to the square of the length of the longest side (hypotenuse).

Note: A right triangle is a triangle that has a right angle (angle with 90 degrees). The side opposite the right angle is the hypotenuse.

Pythagorean Theorem: $a^2 + b^2 = c^2$

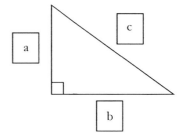

Example:

Find the length of the missing side of the right triangle.

Solution:

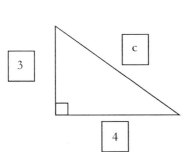

$$a^2 + b^2 = c^2$$
$$(3)^2 + (4)^2 = c^2$$
$$9 + 16 = c^2$$
$$25 = c^2$$
$$c = -5 \text{ or } +5$$

Note: $c^2 = 25$
$c = -5$ because $(-5)^2 = 25$
$c = 5$ because $(5)^2 = 25$

Note: Since we are looking for a distance, the length can not be negative.

$$c = 5$$

Examples:

Find the length of the missing side of the right triangles.

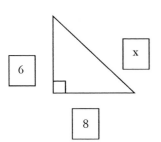 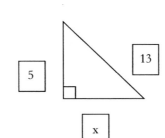

CHAPTER 6

Distance Formula - the distance d between the points with coordinates (x_1, y_1) and (x_2, y_2)

$$d = \sqrt{(x_2 - x_1)^2 + (y_2 - y_1)^2}$$

Example:

Find the distance between: (-3, 4) and (2, 5) using the distance formula.

SOLUTION:

First match up (-3, 4) and (2, 5) with (x_1, y_1) and (x_2, y_2)

$(x_1 = -3, y_1 = 4)$ and $(x_2 = 2, y_2 = 5)$

Then substitute those values into the distance formula:

$$d = \sqrt{(x_2 - x_1)^2 + (y_2 - y_1)^2}$$
$$d = \sqrt{(2 - (-3))^2 + (5 - 4)^2}$$
$$d = \sqrt{(5)^2 + (1)^2}$$
$$d = \sqrt{25 + 1} = \sqrt{26} \approx 5.1$$

Example:

Find the distance between: $(2, -7)$ and $(-4, 1)$

What are square roots and when are they used?

_____

_____

6.1: AN INTRODUCTION TO SQUARE ROOTS - PRACTICE PROBLEMS

Simplify:

1. $\sqrt{9}$ _____
2. $\sqrt{36}$ _____
3. $\sqrt{100}$ _____

4. $\sqrt{144}$ _____
5. $\sqrt{625}$ _____
6. $\sqrt{10,000}$ _____

Estimate the following square roots:

7. $\sqrt{3}$ _____
8. $\sqrt{30}$ _____

9. $\sqrt{51}$ _____
10. $\sqrt{92}$ _____

Simplify:

11. $\sqrt{x^6}$ _____
12. $\sqrt{x^{30}}$ _____

252 CHAPTER 6

Simplify:

13. $\sqrt{4x^2}$ _____

14. $\sqrt{81x^4y^{12}}$ _____

15. $\sqrt{100x^8y^{18}z^{10}}$ _____

16. $\sqrt{144x^{50}y^{200}}$ _____

Find the length of the missing side:

17.

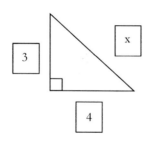

18.

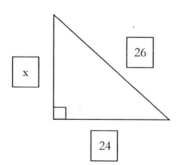

19. Find the distance between

   $(-2, 9)$ and $(1, 5)$

   $d = \sqrt{(x_2 - x_1)^2 + (y_2 - y_1)^2}$

20. Find the distance between

   $(-3, 5)$ and $(-6, -2)$

   $d = \sqrt{(x_2 - x_1)^2 + (y_2 - y_1)^2}$

6.2: SIMPLIFY SQUARE ROOTS

How do you simplify square roots?

Simplify Square Roots

Simplify $\sqrt{18}$
Method 1: Factor the radicand into prime factors and find pairs
Step 1: Factor the radicand into prime factors $\sqrt{18}$ $\sqrt{3 \cdot 3 \cdot 2}$
Step 2: Circle the pairs $\sqrt{(3 \cdot 3) \cdot 2}$
Step 3: For every pair one number comes out of the radical $3\sqrt{2}$

Simplify $\sqrt{18}$
Method 2: Find perfect squares
Step 1: Determine 2 numbers that multiply to 18, where one of the numbers is a perfect square. Then separate the numbers into 2 radicals. $\sqrt{18}$ $\sqrt{9}\sqrt{2}$
Step 2: Take the square root of the perfect square. $\sqrt{9}\sqrt{2}$ $3\sqrt{2}$

Simplify completely the following square roots

1. $\sqrt{12}$

2. $\sqrt{200}$

3. $\sqrt{72}$

4. $5\sqrt{20}$

5. $\sqrt{32}$

6. $\sqrt{150}$

RADICALS

Simplify $\sqrt{x^9}$
Method 1: Factor the radicand into prime factors and find pairs
Step 1: Factor the radicand into prime factors $$\sqrt{x^9}$$ $$\sqrt{x \cdot x \cdot x \cdot x \cdot x \cdot x \cdot x \cdot x \cdot x}$$
Step 2: Circle the pairs $$\sqrt{(x \cdot x) \cdot (x \cdot x) \cdot (x \cdot x) \cdot (x \cdot x) \cdot x}$$
Step 3: For every pair one number comes out of the radical $$x^4 \sqrt{x}$$

Simplify $\sqrt{x^9}$
Method 2: Find perfect squares
Step 1: Separate the radicand into perfect squares and non perfect squares. $$\sqrt{x^9}$$ $$\sqrt{x^8} \sqrt{x}$$
Step 2: Take the square root of the perfect squares. $$\sqrt{x^8} \sqrt{x}$$ $$x^4 \sqrt{x}$$

Examples:

Simplify

1. $\sqrt{x^{11}}$
2. $\sqrt{x^6 y^{15}}$
3. $\sqrt{x^{81}}$

Simplify $\sqrt{20x^5y^2}$

Method 1: Factor the radicand into prime factors and find pairs.

Step 1: Factor radicand into prime factors

$\sqrt{20x^5y^2}$

$\sqrt{2 \cdot 2 \cdot 5 \cdot x \cdot x \cdot x \cdot x \cdot x \cdot y \cdot y}$

Step 2: Circle the pairs

$\sqrt{(2 \cdot 2) \cdot 5 \cdot (x \cdot x) \cdot (x \cdot x) \cdot x \cdot (y \cdot y)}$

Step 3: For every pair one number comes out of the radical

$2x^2 y\sqrt{5x}$

Simplify $\sqrt{20x^5y^2}$

Method 2: Find perfect squares

Step 1: Separate the radicand into perfect squares and non perfect squares.

$\sqrt{20x^5y^2}$

$\sqrt{4x^4y^2} \cdot \sqrt{5x}$

Step 2: Take the square root of the perfect squares.

$\sqrt{4x^4y^2} \cdot \sqrt{5x}$

$2x^2 y\sqrt{5x}$

Examples:

Simplify

1. $\sqrt{12x^5}$

2. $\sqrt{100y^3}$

3. $5\sqrt{45x^5y^{12}}$

4. $2\sqrt{25x^{10}y}$

5. $3x^2\sqrt{15x^7y^2}$

6. $2x^2y\sqrt{50x^{14}y^5z}$

Quotient Rule to Simplify Square Roots

The square root of the quotient of two numbers is equal to the quotient of their square roots.

$$\sqrt{\frac{a}{b}} = \frac{\sqrt{a}}{\sqrt{b}}$$

1. Simplify: $\sqrt{\dfrac{9}{16}}$

2. Simplify: $\sqrt{\dfrac{101}{25}}$

3. Simplify: $\sqrt{\dfrac{50a^5 b^7}{8ab^{10}}}$

How do we simplify square roots?

6.2: SIMPLIFY SQUARE ROOTS - PRACTICE PROBLEMS

Simplify:

1. $\sqrt{28}$ _____
2. $\sqrt{54}$ _____
3. $2\sqrt{125}$ _____

4. $\sqrt{48}$ _____
5. $3\sqrt{162}$ _____
6. $\sqrt{108}$ _____

7. $\sqrt{x^3}$ _____
8. $\sqrt{x^9}$ _____
9. $\sqrt{x^6 y^5}$ _____

10. $x\sqrt{x^7}$ _____
11. $x^2\sqrt{x^{10}}$ _____
12. $x^3 y\sqrt{x^{11} y^{14}}$ _____

CHAPTER 6

Simplify:

13. $\sqrt{40x^3}$ _____ 14. $\sqrt{49y^{11}}$ _____

15. $2\sqrt{50x^6y^{21}}$ _____ 16. $3x\sqrt{81x^{25}y^{100}}$ _____

17. $4xy\sqrt{45x^{16}y^{27}}$ _____ 18. $5x^2y^3\sqrt{72x^3y^8z}$ _____

19. $\sqrt{\dfrac{36}{121}}$ _____ 20. $\sqrt{\dfrac{27a^6b^5}{20a^3b^{15}}}$ _____

6.3: ADDING AND SUBTRACTING RADICAL EXPRESSIONS

How do we add and subtract radical expressions?

Review Problems:

Simplify

1. $3x + 4x$
2. $6x - 3y - 4x + 8y$
3. $2x^2 - 3x + 5x^2$

Like Radicals: Square root radicals are called like radicals when they have the same radicand.

1. Like Radicals $5\sqrt{3} + 2\sqrt{3}$
2. Unlike Radicals $2\sqrt{5} + 7\sqrt{6}$

Combining Like Radicals

1. $3\sqrt{2} + 5\sqrt{2}$
2. $2\sqrt{3} + 4\sqrt{5} - 6\sqrt{3} + \sqrt{5}$
3. $2\sqrt{x} + 7\sqrt{x}$

4. $7\sqrt{3x} + 2\sqrt{3x}$
5. $7\sqrt{ab} + 2\sqrt{a^2b} - 3\sqrt{ab}$
6. $5 + 3\sqrt{x} + 4x$

Simplifying Radicals

Example:

$$\sqrt{50} + \sqrt{32}$$

SOLUTION:

Step 1: Simplify the radicals

$\sqrt{50} + \sqrt{32}$
$\sqrt{25} \cdot \sqrt{2} + \sqrt{16} \cdot \sqrt{2}$
$5\sqrt{2} + 4\sqrt{2}$

Step 2: Combine like radicals

$5\sqrt{2} + 4\sqrt{2}$
$9\sqrt{2}$

Examples:

1. $\sqrt{45} - \sqrt{27}$

2. $5\sqrt{8} + 2\sqrt{18}$

3. $2\sqrt{24x^2} - x\sqrt{54}$

4. $2\sqrt{32x} - 6\sqrt{5y} + 5\sqrt{200x} + 3\sqrt{125y}$

How do we add and subtract radical expressions?

_____

_____

6.3: ADDING AND SUBTRACTING RADICAL EXPRESSIONS - PRACTICE PROBLEMS

Solve:

1. $4\sqrt{3} + 7\sqrt{3}$

2. $5\sqrt{2} + \sqrt{7} - 6\sqrt{2} + 8\sqrt{7}$

3. $5\sqrt{xy} + 4\sqrt{x} - 8\sqrt{xy}$

4. $5x\sqrt{y} + 9\sqrt{3x} - 8\sqrt{y} + 6\sqrt{3x}$

5. $-4x\sqrt{a^2b^3} + 2\sqrt{a^2b^3} - 6x\sqrt{a^2b^3}$

6. $\sqrt{12} - \sqrt{27}$

7. $\sqrt{32} + \sqrt{18}$

8. $\sqrt{27} - \sqrt{48} + \sqrt{75}$

9. $3\sqrt{12} + 2\sqrt{48}$

10. $\sqrt{108} + 3\sqrt{75} - 2\sqrt{50}$

11. $3\sqrt{8} + \sqrt{48} - 5\sqrt{27} - 2\sqrt{32}$

12. $5\sqrt{54x^2} - 2\sqrt{24x^2}$

13. $2a\sqrt{48ab^2} - b\sqrt{27a^3} + 3\sqrt{75a^3b^2}$

6.4: MULTIPLYING AND DIVIDING RADICAL EXPRESSIONS

How do we multiply and divide radical expressions?

Product rule to simplify square roots:

$$\sqrt[n]{a} \cdot \sqrt[n]{b} = \sqrt[n]{ab}$$

Example:

$4\sqrt{6} \cdot 5\sqrt{3}$

Solution:

Step 1: To multiply radical expressions, multiply the coefficients, multiply the radicands

$4\sqrt{6} \cdot 5\sqrt{3}$

$4 \cdot 5 \sqrt{6 \cdot 3}$

$20\sqrt{18}$

Step 2: If needed, then simplify the radicand

$20\sqrt{18}$

$20\sqrt{9}\sqrt{2}$

$20 \cdot 3\sqrt{2}$

$60\sqrt{2}$

Examples:

1. $\sqrt{5} \cdot 3\sqrt{2}$

   $3\sqrt{10}$

2. $\sqrt{6x} \cdot \sqrt{8x}$

   $\sqrt{48x^2}$

   $\sqrt{16 \cdot 3 x^2}$

   $4x\sqrt{3}$

CHAPTER 6

3. $2\sqrt{5x^3} \cdot 3\sqrt{4x^6}$
   $6\sqrt{20x^9}$
   $6\sqrt{5 \cdot 4x^9}$
   $12x^4\sqrt{5x}$

4. $4\sqrt{2x} \cdot 3\sqrt{14y}$
   $12\sqrt{2x \cdot 14y}$
   $12\sqrt{28xy}$
   $12\sqrt{7 \cdot 4xy}$
   $24\sqrt{7xy}$

Square of a square root: $(\sqrt{x})^2 = x$, for any positive real number x

Examples:

1. $(\sqrt{5})^2$
   5

2. $(\sqrt{x-3})^2$
   $x-3$

3. $(3\sqrt{2})^2$
   $3 \cdot 2 = 6$

Multiplying Radicals Expressions

Example:

$3\sqrt{2x}(5\sqrt{4x^3} - \sqrt{3x^2})$

Solution:

Step 1: Use the distributive property

$3\sqrt{2x}(5\sqrt{4x^3} - \sqrt{3x^2})$
$3\sqrt{2x}(5\sqrt{4x^3}) + 3\sqrt{2x}(-\sqrt{3x^2})$

Step 2: Simplify the radicals

$15\sqrt{8x^4} \quad -3\sqrt{6x^3}$
$15\sqrt{4x^4} \cdot \sqrt{2} - 3\sqrt{x^2}\sqrt{6x}$
$15 \cdot 2x^2\sqrt{2} \quad -3x\sqrt{6x}$
$30x^2\sqrt{2} - 3x\sqrt{6x}$

RADICALS

Example:

$(3\sqrt{2} - 4)(\sqrt{2} + 3)$

SOLUTION:

Step 1: Use the distributive property or FOIL

$(3\sqrt{2} - 4)(\sqrt{2} + 3)$

$3\sqrt{2}(\sqrt{2}) + 3\sqrt{2}(3) - 4(\sqrt{2}) - 4(3)$

Step 2: Multiply the radicals

$3\sqrt{2}(\sqrt{2}) + 3\sqrt{2}(3) - 4(\sqrt{2}) - 4(3)$

$3\sqrt{4} + 9\sqrt{2} - 4\sqrt{2} - 12$

Step 3: Simplify the radicals

$3\sqrt{4} + 9\sqrt{2} - 4\sqrt{2} - 12$

$3(2) + 9\sqrt{2} - 4\sqrt{2} - 12$

$6 + 5\sqrt{2} - 12$

$-6 + 5\sqrt{2}$

Examples:

1. $4\sqrt{5}(3\sqrt{5} - 6)$

   $(4\sqrt{5} \cdot 3\sqrt{5}) + (4\sqrt{5} \cdot 6)$

   $12\sqrt{25} - 24\sqrt{5}$

   $60 - 24\sqrt{5}$

2. $\sqrt{2}(7\sqrt{9} - \sqrt{8})$

   $7\sqrt{18} - \sqrt{16}$

   $7\sqrt{9 \cdot 2} - \sqrt{8 \cdot 2}$

   $21\sqrt{2} -$

3. $2\sqrt{3x}(\sqrt{4x} - 3\sqrt{x^2})$

Examples:

1. $(2\sqrt{3}+5)(\sqrt{5}+2)$
$2\sqrt{15} + 4\sqrt{3} + 5\sqrt{5} + 10$

2. $(3\sqrt{6}-2)(2\sqrt{6}+5)$
$6\sqrt{36} + 15\sqrt{6} - 4\sqrt{6} - 10$
$36 + 11\sqrt{6} - 10$
$\boxed{26 + 11\sqrt{6}}$

3. $(\sqrt{7}-3)^2$
$(\sqrt{7}-3)(\sqrt{7}-3)$
$\sqrt{49} - 3\sqrt{7} - 3\sqrt{7} + 9$
$7 - 6\sqrt{7} + 9$
$16 - 6\sqrt{7}$

4. $(\sqrt{3x}-2)(4\sqrt{3x}-6)$
$4\sqrt{9x^2} - 6\sqrt{3x} - 8\sqrt{3x} + 12$
$12x - 14\sqrt{3x} + 12$

Dividing Radicals Expressions

Examples:

$\dfrac{\sqrt{40}}{\sqrt{5}} =$

Since they are both have a radical we can rewrite as:

$\sqrt{\dfrac{40}{5}} =$

Then we can divide:

$\sqrt{8} =$

Then we can simplify:

$\sqrt{4} \cdot \sqrt{2} =$
$2\sqrt{2}$

$\dfrac{\sqrt{54x^4}}{\sqrt{2x}} =$

Since they are both have a radical we can rewrite as :

$\sqrt{\dfrac{54x^4}{2x}} =$

Then we can divide:

$\sqrt{27x^3} =$

Then we can simplify:

$\sqrt{9x^2} \cdot \sqrt{3x} =$
$3x\sqrt{3x}$

Examples:

1. $\dfrac{\sqrt{40}}{\sqrt{5}} = \sqrt{\dfrac{40}{5}} = \sqrt{8} = \sqrt{2 \cdot 4} = 2\sqrt{2}$

2. $\dfrac{\sqrt{84}}{\sqrt{3}} = \sqrt{\dfrac{84}{3}} = \sqrt{28} = \sqrt{4 \cdot 7} = 2\sqrt{7}$

3. $\dfrac{\sqrt{32x^6}}{\sqrt{2x^2}} = \sqrt{16x^4} = 4x^2$

4. $\dfrac{\sqrt{200x}}{\sqrt{4x}} \quad \sqrt{50} = \sqrt{25 \cdot 2} = 5\sqrt{2}$

Applications

Find the area of the rectangle.

$4\sqrt{6}$ ft

$5\sqrt{3}$ ft

Find the area of the rectangle.

$2\sqrt{3} + 5$ ft

$4\sqrt{3} - 6$ ft

How do we multiply and divide radical expressions?

6.4: MULTIPLYING AND DIVIDING RADICAL EXPRESSIONS PRACTICE PROBLEMS

Simplify

1. $\sqrt{3}(4\sqrt{2}-5)$

2. $3\sqrt{5}(4\sqrt{2}-\sqrt{6})$

3. $5\sqrt{2}(2\sqrt{8}-7)$

4. $3\sqrt{10}(4\sqrt{2}-6\sqrt{5})$

5. $2\sqrt{2x}(\sqrt{14x}-6\sqrt{x^5})$

6. $x\sqrt{3}(x\sqrt{8x^2y}-5xy\sqrt{3x^3y^2})$

7. $(2\sqrt{3}+4)(3\sqrt{2}+5)$ — foil!
   $6\sqrt{6} + 10\sqrt{3} + 12\sqrt{2} + 20$

8. $(5\sqrt{2}+3)(3\sqrt{2}-4)$
   $15\sqrt{4} - 20\sqrt{2}$

9. $(\sqrt{3}+5)^2$

$(\sqrt{3}+5)(\sqrt{3}+5)$
$\sqrt{9}+5\sqrt{3}+5\sqrt{3}+25$
$\sqrt{9}+10\sqrt{3}+25$
$3+10\sqrt{3}+25$
$28+10\sqrt{3}$

10. $(2\sqrt{6}-7)^2$

$(2\sqrt{6}-7)(2\sqrt{6}-7)$
$4\sqrt{36}-14\sqrt{6}-14\sqrt{6}+49$
$4\sqrt{36}-28\sqrt{6}+49$

11. $(4\sqrt{3x}+3x)(3\sqrt{15x}-\sqrt{3x})$

$12\sqrt{3x \cdot 15x} - 4\sqrt{9x^2} + 9x\sqrt{15x} - 3x\sqrt{3x}$
$36x\sqrt{5} - 12x + 9x\sqrt{15x} - 3x\sqrt{3x}$

12. $(5x\sqrt{6xy}-\sqrt{2y})(\sqrt{3x^3}-2x^2\sqrt{2xy})$

13. $\dfrac{\sqrt{54}}{\sqrt{3}} = \sqrt{\dfrac{54}{3}}$

$= \sqrt{18}$
$= \sqrt{9 \cdot 2}$
$= 3\sqrt{2}$

14. $\dfrac{\sqrt{90x^{10}}}{\sqrt{2x^3}}$

$= \sqrt{\dfrac{90x^{10}}{2x^3}}$
$= \sqrt{45x^7}$
$= 3x^3\sqrt{5x}$

15. Find the area of square.

$5\sqrt{2}-4\,ft$

6.5: RATIONALIZING THE DENOMINATOR (MONOMIALS ONLY)

How do we Rationalize the Denominator?

<u>Remember this</u>: **Pythagorean Theorem-** In a right triangle, the sum of the squares of the lengths of the two shorter sides (legs) is equal to the square of the length of the longest side (hypotenuse).

Pythagorean Theorem: $a^2 + b^2 = c^2$

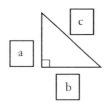

Practice Example:

Find the length of the missing side. 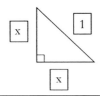
SOLUTION: $a^2 + b^2 = c^2$ $x^2 + x^2 = 1^2$ $2x^2 = 1$ $\dfrac{2x^2}{2} = \dfrac{1}{2}$ $x^2 = \dfrac{1}{2}$ $x^2 = \dfrac{1}{2}$ $x = -\sqrt{\dfrac{1}{2}}$ or $+\sqrt{\dfrac{1}{2}}$ Note: Since we are looking for a distance, the length cannot be negative. $x = \sqrt{\dfrac{1}{2}} = \dfrac{\sqrt{1}}{\sqrt{2}} = \dfrac{1}{\sqrt{2}}$
Since the denominator is an irrational number we need to change the denominator to a rational number by rationalizing the denominator.

Steps to rationalize the denominator

1. Multiply the numerator and denominator by the square root in the denominator

$$\frac{1}{\sqrt{2}} \cdot \frac{\sqrt{2}}{\sqrt{2}} = \frac{\sqrt{2}}{\sqrt{2 \cdot 2}} = \frac{\sqrt{2}}{\sqrt{4}}$$

2. Simplify

$$\frac{\sqrt{2}}{\sqrt{4}} = \frac{\sqrt{2}}{2}$$

Examples:

Rationalize the denominator: $\dfrac{7}{\sqrt{6}}$	Rationalize the denominator: $\dfrac{3}{\sqrt{5x}}$
SOLUTION: Rationalize the denominator: $\dfrac{7}{\sqrt{6}} \cdot \dfrac{\sqrt{6}}{\sqrt{6}} = \dfrac{7\sqrt{6}}{\sqrt{36}} = \dfrac{7\sqrt{6}}{6}$	**SOLUTION:** Rationalize the denominator: $\dfrac{3}{\sqrt{5x}} \cdot \dfrac{\sqrt{5x}}{\sqrt{5x}} = \dfrac{3\sqrt{5x}}{\sqrt{25x^2}} = \dfrac{3\sqrt{5x}}{5x}$

RADICALS 275

Examples:

Simplify completely

1. Rationalize the denominator:

    $$\frac{3}{\sqrt{7}}$$

2. Rationalize the denominator:

    $$\frac{4}{\sqrt{11}} = \frac{4}{\sqrt{11}} \cdot \frac{\sqrt{11}}{\sqrt{11}} = \boxed{\frac{4\sqrt{11}}{11}}$$

3. Rationalize the denominator:

    $$\frac{\sqrt{2}}{\sqrt{3}} = \frac{\sqrt{2}}{\sqrt{3}} \cdot \frac{\sqrt{3}}{\sqrt{3}} = \frac{\sqrt{6}}{3}$$

4. Rationalize the denominator:

    $$\frac{9}{\sqrt{2y}} = \frac{9}{\sqrt{2y}} \cdot \frac{\sqrt{2y}}{\sqrt{2y}} = \frac{9\sqrt{2y}}{2y}$$

5. Rationalize the denominator:

    $$\frac{2x}{\sqrt{10x}} = \frac{2x}{\sqrt{10x}} \cdot \frac{\sqrt{10x}}{\sqrt{10x}} = \frac{2x\sqrt{10x}}{10x}$$
    $$= \frac{\sqrt{10x}}{5}$$

6. Rationalize the denominator:

    $$\frac{6}{\sqrt{5xy}} \cdot \frac{\sqrt{5xy}}{\sqrt{5xy}} = \frac{6\sqrt{5xy}}{5xy}$$

How do we Rationalize the Denominator?

_____

_____

More Examples:

Simplify completely

7. Rationalize the denominator:

$$\frac{1}{\sqrt{7}} \cdot \frac{\sqrt{7}}{\sqrt{7}} = \frac{\sqrt{7}}{7}$$

8. Rationalize the denominator:

$$\frac{2}{\sqrt{12}} = \frac{2\sqrt{12}}{12} = \frac{\sqrt{12}}{6}$$

$$= \frac{\sqrt{3 \cdot 4}}{6} = \frac{2\sqrt{3}}{6} = \frac{\sqrt{3}}{3}$$

9. Rationalize the denominator:

$$\frac{\sqrt{12}}{\sqrt{3}} = \frac{\sqrt{4 \cdot 3}}{\sqrt{3}} = \frac{2\sqrt{3}}{\sqrt{3}}$$

$$= 2$$

OR

$$\sqrt{\frac{12}{3}} = \sqrt{4} = 2$$

10. Rationalize the denominator:

$$\frac{16}{\sqrt{6z}} \cdot \frac{\sqrt{6z}}{\sqrt{6z}} = \frac{16\sqrt{6z}}{6z}$$

$$= \frac{8\sqrt{6z}}{3z}$$

11. Rationalize the denominator:

$$\frac{3x^2}{\sqrt{2x}}$$

12. Rationalize the denominator:

$$\frac{6}{\sqrt{7xyz}}$$

6.6: SOLVING RADICAL EQUATIONS

How do we solve radical equations?

Example:

Solve for x: $3\sqrt{x+2}+5=14$

SOLUTION:

Step 1: Isolate the radical

$$3\sqrt{x+2}+5=14$$
$$\phantom{3\sqrt{x+2}}-5-5$$
$$3\sqrt{x+2}=9$$
$$\div 3 \phantom{3\sqrt{x+2}=}\div 3$$
$$\sqrt{x+2}=3$$

Step 2: Remove the radical by squaring both sides

$$\sqrt{x+2}=3$$
$$\left(\sqrt{x+2}\right)^2=(3)^2$$
$$x+2=9$$

Step 3: Solve for x

$$x+2=9$$
$$-2-2$$
$$x=7$$

Step 4: Check

$$3\sqrt{x+2}+5=14$$
$$3\sqrt{7+2}+5=14$$
$$3\sqrt{9}+5=14$$
$$3 \cdot 3+5=14$$
$$9+5=14$$
$$14=14$$

Examples:

1. Solve for x: $\sqrt{x-5} - 3 = 4$

2. Solve for x: $\sqrt{x-2} + 5 = 2$

3. Solve for x: $\sqrt{x+4} - x = -2$

4. Solve for x: $\sqrt{x-4} = 2\sqrt{x-16}$

Applications

The perimeter of the triangle is 13 feet.
Solve for x, and then find the missing side.

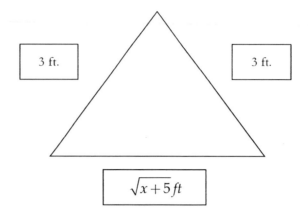

The perimeter of the rectangle is 22 feet.
Solve for x, and then find the length.

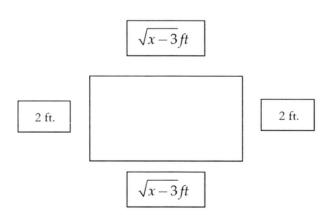

How do we solve radical expressions?

_____

_____

6.6: SOLVING RADICAL EQUATIONS - PRACTICE PROBLEMS

Solve:

1. $\sqrt{x} = 9$
2. $\sqrt{x} = -2$
3. $\sqrt{2x} = 4$

4. $\sqrt{x} + 3 = 7$
5. $\sqrt{x+2} = 3$
6. $\sqrt{x-7} = 10$

7. $\sqrt{2x-4} = 6$
8. $\sqrt{x+1} - 6 = 3$
9. $\sqrt{3x-14} + 3 = 7$

10. $\sqrt{x^2 + 2x - 6} = x$

11. $\sqrt{x+3} = x+1$

12. $\sqrt{5x+6} = \sqrt{7x-6}$

13. $2\sqrt{3x+4} = \sqrt{5x+9}$

14. The perimeter of the rectangle is 18 feet. Solve for x, and then find the length.

$\sqrt{4x-8}$ ft

3 ft.

3 ft.

$\sqrt{4x-8}$ ft

6.7: HIGHER ORDER ROOTS

what are and how do we solve higher order roots?

Higher Order Roots

Square roots are not the only roots. There are many different types of roots. Here are some examples:

$\sqrt[3]{125} = \sqrt[3]{5 \cdot 5 \cdot 5} = 5$

$\sqrt[4]{81} = \sqrt[4]{3 \cdot 3 \cdot 3 \cdot 3} = 3$

$\sqrt[5]{32} = \sqrt[5]{2 \cdot 2 \cdot 2 \cdot 2 \cdot 2} = 2$

Simplify the following roots:

Examples:

1. $\sqrt[3]{27}$

2. $\sqrt[4]{16}$

3. $\sqrt[3]{x^6}$

4. $\sqrt[5]{x^{15} y^{25}}$

Examples:

Simplify

1. $\sqrt[3]{40}$

2. $\sqrt[4]{80}$

3. $\sqrt[3]{-128x^6y^5}$

4. $\sqrt[4]{-32}$

5. $\sqrt[3]{54x^{10}y^{12}}$

6. $\sqrt[3]{\dfrac{32}{125}}$

What are and how do we solve higher order roots?

_____

_____

6.7: HIGHER ORDER ROOTS PRACTICE PROBLEMS

Simplify:

1. $\sqrt[3]{64}$

2. $\sqrt[4]{81}$

3. $\sqrt[5]{-32}$

4. $\sqrt[3]{x^{12}}$

5. $\sqrt[4]{x^4 y^{16} z^{24}}$

6. $\sqrt[6]{a^{12} b^{30}}$

7. $\sqrt[3]{24}$

8. $\sqrt[3]{32}$

9. $\sqrt[4]{256}$

10. $\sqrt[3]{72 x^9 y^7}$

11. $\sqrt[4]{48 x^{11} y^{40} z}$

12. $\sqrt[3]{250 x^5 y^{30} z^3}$

13. $\sqrt[3]{\dfrac{54}{27}}$

MAT0028C DEVELOPMENTAL MATH II Name: _____

TEST 4 (CHAPTER 6) REVIEW

Questions from Chapter 6

1. Simplify: $\sqrt{121}$

2. Simplify: $\sqrt{20}$

3. Simplify: $\sqrt{x^7 y^{10}}$

4. Simplify: $\sqrt{63x^2 y^9 z}$

5. Simplify: $5\sqrt{49x^4 y^{11}}$

6. Simplify: $3x^2 \sqrt{200xy^{20} z^3}$

7. Simplify: $3\sqrt{5} \cdot \sqrt{8}$

8. Simplify: $2\sqrt{3}\left(5\sqrt{3} - \sqrt{5}\right)$

9. Simplify: $4\sqrt{2x}\left(\sqrt{3x^3} - 2\sqrt{4x^4}\right)$

10. Simplify: $\sqrt{5}\left(2\sqrt{5} - 6\right)$

11. Simplify: $\left(4\sqrt{3} + 6\right)\left(2\sqrt{3} - 5\right)$

12. Simplify: $\left(4\sqrt{3} - 2\right)^2$

13. Rationalize the denominator

$\dfrac{5\sqrt{3}}{\sqrt{2}}$

14. Rationalize the denominator

$\dfrac{3\sqrt{2x}}{\sqrt{3x}}$

15. Simplify: $\sqrt[4]{10{,}000}$

16. Simplify: $\sqrt[3]{16x^6 y^5}$

Review Questions

17. Factor: $30a^{10}b^5 + 20a^5b^6 - 5a^2b^3$

18. Factor: $3xy + 3xz - 5y - 5z$

19. Factor: $x^2 + 5x - 24$

20. Factor: $10x^2 + 21x - 10$

21. Factor: $81x^4 - 49y^6$

22. Solve: $x^2 - x - 42 = 0$

23. Solve: $4x^2 + 8x + 3 = 0$

24. Simplify: $\dfrac{2x^2 - 5x + 3}{x^2 - 1}$

25. Simplify: $|4|-|-3|+|2-8|$

26. Solve for b: $5c = 6a - 2b$

27. Find the x–intercept for: $-x + 7y = 12$

28. Graph: $y = -\dfrac{2}{3}x + 5$

29. Solve: $10 - 2x \geq 2(2x - 3)$

30. Simplify: $\dfrac{\left(a^2 b^3\right)^4 \left(ab^6\right)}{a^{10} b^5}$

31. Translate into an equation: The square of a number less than 4 times a number is 5 more than twice a number.

32. The length of a rectangular pool is 5 less than twice the width. The perimeter of the pool is 110 feet. Find the length and width of the pool. Label each distance correctly.

32. If a bed cost $675 after a 20% discount, what was the original cost?

MAT0028C
DEVELOPMENTAL MATH II
PRACTICE PROBLEMS

PRACTICE PROBLEMS

MAT0028C DEVELOPMENTAL MATH II STATE OF FLORIDA COMPETENCY LIST

COMPETENCY ID	MATHEMATICS CATEGORY	MATHEMATICS COMPETENCIES – UPPER	CONNECTION TO TEXTBOOK
MDECU1	Exponents & Polynomials	Applies the order of operations to evaluate algebraic expressions, including those with parentheses and exponents	Chapter 4 throughout
MDECU2	Exponents & Polynomials	Simplifies an expression with integer exponents	4.1
MDECU3	Exponents & Polynomials	Add, subtract, multiply, and divide polynomials. Division by monomials only. *(Does not include division by binomials)*	4.5, 4.6, 4.7
MDECU4	Factoring	Solve quadratic equations in one variable by factoring	5.6
MDECU5	Factoring	Factor polynomial expressions (GCF, grouping, trinomials, difference of squares)	5.1, 5.2, 5.3, 5.4, 5.5
MDECU6	Graphing	Graph linear equations using table of values, intercepts, slope intercept form	3.2, 3.3, 3.5
MDECU7	Linear Equations	Solve linear equations in one variable using manipulations guided by the rules of arithmetic and the properties of equality.	2.2
MDECU8	Linear Equations	Solve literal equations for a given variable with applications (geometry, motion [d=rt], simple interest [i=prt])	2.5
MDECU9	Radicals	Simplify radical expressions – square roots only	6.1, 6.2
MDECU10	Radicals	Adds, subtracts, and multiplies square roots of monomials	6.3, 6.4
MDECU11	Exponents & Polynomials	Convert between scientific notation and standard notation	4.3
MDECU12	Exponents & Polynomials	Solve application problems involving geometry (perimeter and area with algebraic expressions)	2.7
MDECU13	Graphing	Identifies the intercepts of a linear equation	3.3, 3.5
MDECU14	Graphing	Identify the slope of a line (from slope formula, graph, and equation)	3.4, 3.5
MDECU15	Linear Equations	Solve multi-step problems involving fractions and percentages (Include situations such as simple interest, tax, markups/markdowns, gratuities and commissions, fees, percent increase or decrease, percent error, expressing rent as a percentage of take-home pay)	2.7
MDECU16	Linear Equations	Solve linear inequalities in one variable and graph the solution set on a number line	2.6
MDECU17	Radicals	Rationalize the denominator (monomials only)	6.5
MDECU18	Radicals	Solve application problems involving geometry (Pythagorean Theorem)	6.1
MDECU19	Rationals	Recognize proportional relationships and solve problems involving rates and ratios	2.3
MDECU20	Rationals	Simplify, multiply, and divide rational expressions	5.7, 5.9
MDECU21	Rationals	Add and subtract rational expressions with monomial denominators	5.10
MDECU22	Rationals	Convert units of measurement across measurement systems	2.4

ORDER OF OPERATIONS (NO GROUPING / NO EXPONENTS)

1. Simplify: $6 - 18 \div 9 - 4$

A. 19 B. 34 C. 7 D. 0

SOLUTION:

Simplify: $6 - 18 \div 9 - 4$

You need to know the order of operations:

1. Parentheses
2. Exponents
3. Multiplication/ Division (whichever comes first left to right)
4. Addition / Subtraction (whichever comes first left to right)

In this problem, the first step would be the division:

$6 - 18 \div 9 - 4$

$6 - 2 - 4$

Then complete the subtraction from left to right.

$6 - 2 - 4$

$4 - 4$

0

The solution is: 0

296 PRACTICE PROBLEMS

ORDER OF OPERATIONS (NO GROUPING / NO EXPONENTS)

1. Simplify: $8 - 7 + 10 \cdot 3$
   a. -29
   b. 33
   c. 31
   d. -43

2. Simplify: $13 - 16 \div 8 + 12$
   a. $-\dfrac{3}{20}$
   b. 27
   c. 23
   d. $\dfrac{61}{5}$

3. Simplify: $7 + 5 \cdot 8 \div 4 \cdot 3 - 6$
   a. 31
   b. -72
   c. $\dfrac{13}{3}$
   d. 66

4. Simplify: $8 - 12 \div 2 \cdot 3 + 5$
   a. 11
   b. 23
   c. -1
   d. -5

SOLUTIONS:

1. c   2. c   3. a   4. d

ORDER OF OPERATIONS (WITH GROUPING AND EXPONENTS)

2. Simplify: $23 - (8)^2 \div (14 - 6) \cdot 6$

A. -25 B. 90 C. $-\dfrac{123}{4}$ D. $\dfrac{65}{3}$

SOLUTION:

Simplify: $23 - (8)^2 \div (14 - 6) \cdot 6$

You need to know the order of operations:

1. Parentheses
2. Exponents
3. Multiplication/ Division (whichever comes first, left to right)
4. Addition / Subtraction (whichever comes first, left to right)

In this problem, the first step would be the parenthesis:

$23 - (8)^2 \div (14 - 6) \cdot 6$

$23 - (8)^2 \div (8) \cdot 6$

The next step would be exponents:

$23 - (8)^2 \div (14 - 6) \cdot 6$

$23 - 64 \div (8) \cdot 6$

The next step would be division because it occurs first from left to right:

$23 - (8)^2 \div (14 - 6) \cdot 6$

$23 - 8 \cdot 6$

The next step would be multiplication:

$23 - 8 \cdot 6$

$23 - 48$

The next step would be subtraction:

$23 - 48$

-25

The solution is: -25

ORDER OF OPERATIONS (WITH GROUPING AND EXPONENTS)

1. Simplify: $24 - (-8)^1 \div (6 - 14) \cdot 7$
   a. 224
   b. 35
   c. $\frac{176}{7}$
   d. 80

2. Simplify: $(25 - 15)^2 \div 5$
   a. 4
   b. 20
   c. 80
   d. 2

3. Simplify: $(1 - 10)^2 - (2 + 3)^3$
   a. −33
   b. 66
   c. −44
   d. 206

4. Simplify: $4 + 3(9 - 5)^2 - 10 \div 2 \cdot 5$
   a. 27
   b. 51
   c. 87
   d. 3

Solutions:

1. d
2. b
3. c
4. a

ABSOLUTE VALUE (WITH ADDITION AND SUBTRACTION)

3. Simplify: $|8 + (-14)| + 9$

A. 16   B. 31   C. 3   D. 15

Solution:

Simplify: $|8 + (-14)| + 9$

You need to know the order of operations.

In this problem, the first step is to complete what is inside the absolute value, which functions similarly to the parentheses.

$|-6| + 9$

The next step is to take the absolute value:

$|-6| + 9$

$6 + 9$

The next step is complete the addition:

$6 + 9$

15

The solution is: 15

ABSOLUTE VALUE (WITH ADDITION AND SUBTRACTION)

1. Simplify: $|-13| + |15| - |-9|$
   a. 37
   b. 19
   c. −11
   d. 7

2. Simplify: $-|-9| + |6|$
   a. 15
   b. 3
   c. −3
   d. −15

3. Simplify: $|5 + (-11)| + 9$
   a. 25
   b. 15
   c. −15
   d. 3

4. Simplify: $|-2| - |-5| + |6-14|$
   a. 15
   b. 13
   c. 17
   d. 5

Solutions:

1. b
2. c
3. b
4. d

SIMPLIFY ALGEBRAIC EXPRESSIONS (USING THE DISTRIBUTIVE PROPERTY)

4. Simplify: $-6[6(x+5)+x]$

A. $42x - 180$ B. $30x + 180$ C. $-42x - 180$ D. $42x + 180$

Solution:

Simplify: $-6[6(x+5)+x]$

In this problem, the first step is to simplify within the brackets. Within the brackets, use the distributive property first.

$-6[6(x+5)+x]$
$-6[6x + 6(5) + x]$
$-6[6x + 30 + x]$

The next step is to combine like terms with the brackets.

$-6[6x + 30 + x]$
$-6[6x + 1x + 30]$
$-6[7x + 30]$

The next step is to use distributive property.

$-6[7x + 30]$
$-6(7x) + -6(30)$
$-42x - 180$

The solution is: $-42x - 180$

SIMPLIFY ALGEBRAIC EXPRESSIONS (USING THE DISTRIBUTIVE PROPERTY)

1. Simplify: $-8x - 4(x + 5) + 8$
   a. $-12x - 12$
   b. $-28x + 8$
   c. $-12x + 4$
   d. $-12x - 20$

2. Simplify: $-3[3(r + 5) + r]$
   a. $12r + 45$
   b. $6r + 45$
   c. $-12r - 45$
   d. $-12r + 45$

3. Simplify: $-2(3x - 4) + 5(x + 6)$
   a. $-11x + 38$
   b. $-x + 22$
   c. $-x + 38$
   d. $11x - 22$

4. Simplify: $-2[4(-6d + 3) - d]$
   a. $47d - 24$
   b. $50d - 24$
   c. $-50d + 12$
   d. $-50d - 12$

SOLUTIONS:

1. a
2. c
3. c
4. b

PRACTICE PROBLEMS

EVALUATE AN ALGEBRAIC EXPRESSION

5. Evaluate the given expression $-6w^2 + 5w + 3$ when $w = -4$:

A. -79 B. -119 C. -113 D. -73

SOLUTION:

Evaluate the given expression $-6w^2 + 5w + 3$ when $w = -4$:

In this problem, the first step is to substitute the value into the variable.
$-6w^2 + 5w + 3$
$-6(-4)^2 + 5(-4) + 3$

Then using order of operations, the next step is to complete the exponent.
$-6(-4)^2 + 5(-4) + 3$
$-6(16) + 5(-4) + 3$

The next step is to complete multiplication from left to right.
$-6(16) + 5(-4) + 3$
$-96 + 5(-4) + 3$
$-96 - 20 + 3$

The next step is to perform the subtraction then addition left to right.
$-96 - 20 + 3$
$-116 + 3$
-113

The solution is: -113

EVALUATE AN ALGEBRAIC EXPRESSION

1. Evaluate the expression $3w^2 - 2w - 5$ when w = −4:
   a. 35
   b. 61
   c. 51
   d. 45

2. Evaluate the expression $-4w^2 + 2w + 3$ when w = −3:
   a. −39
   b. −27
   c. −45
   d. −33

3. Evaluate the expression $4xy - z$ when x = −9, y = 4, z = −7:
   a. −137
   b. 151
   c. 137
   d. −151

4. Evaluate the expression $9xy - z^2$ when x = −9, y = 2, z = −2:
   a. −158
   b. 158
   c. −160
   d. −166

Solutions:

1. c
2. a
3. a
4. d

SOLVE A LINEAR EQUATION

6. Solve for r: $-2(-9r - 6) = 6(r + 5)$

   A. $r = \dfrac{7}{2}$ B. $r = -\dfrac{3}{2}$ C. $r = \dfrac{7}{4}$ D. $r = \dfrac{3}{2}$

SOLUTION:

Solve for r: $-2(-9r - 6) = 6(r + 5)$

The first step is to simplify both sides of the equation using distributive property.
$$-2(-9r - 6) = 6(r + 5)$$
$$-2(-9r) + -2(-6) = 6(r) + 6(5)$$
$$18r + 12 = 6r + 30$$

The next step is to move the variables to the same side of the equation.
$$\begin{array}{r} 18r + 12 = 6r + 30 \\ -6r \qquad\quad -6r \\ \hline 12r + 12 = 30 \end{array}$$

The next step is to move the variables to other side of the equation.
$$\begin{array}{r} 12r + 12 = 30 \\ -12 \; -12 \\ \hline 12r = 18 \end{array}$$

The next step is to solve for r and simplify solution.
$$\begin{array}{r} 12r = 18 \\ \div 12 \; \div 12 \end{array}$$
$$r = \dfrac{18}{12} = \dfrac{3}{2}$$

The solution is: $r = \dfrac{3}{2}$

Note: You can check your solution by substituting it back into your equation.

SOLVE A LINEAR EQUATION

1. Solve for r: $9(-2r-7) = -4r+9$

   a. $r = -\dfrac{36}{11}$
   b. $r = \dfrac{27}{11}$
   c. $r = \dfrac{27}{7}$
   d. $r = -\dfrac{36}{7}$

2. Solve for x: $-2(-3x-2) = 8x-3$

   a. $x = \dfrac{7}{2}$
   b. $x = -\dfrac{1}{2}$
   c. $x = \dfrac{1}{14}$
   d. $x = -\dfrac{1}{2}$

3. Solve for y: $-6(3y+4) = 2(y+7)$

   a. $y = \dfrac{19}{10}$
   b. $y = \dfrac{1}{2}$
   c. $y = \dfrac{5}{8}$
   d. $y = -\dfrac{19}{10}$

4. Solve for y: $-2(y+3) = 4(2y-4)$

   a. $y = -1$
   b. $y = -\dfrac{13}{5}$
   c. $y = 1$
   d. $y = \dfrac{5}{3}$

SOLUTIONS:

1. d   2. a   3. d   4. c

SOLVE A LINEAR EQUATION WITH FRACTION COEFFICIENT(S)

7. Solve for y: $\dfrac{5}{6}y + 2 = \dfrac{7}{4}$

A. $y = \dfrac{3}{2}$ B. $y = \dfrac{9}{2}$ C. $y = -\dfrac{5}{24}$ D. $y = -\dfrac{3}{10}$

Solution:

Solve for y: $\dfrac{5}{6}y + 2 = \dfrac{7}{4}$

One way to solve this problem is to eliminate the fractions by multiplying the entire equation by the least common denominator of the fractions. In this case the least common denominator is 12.

$\dfrac{5}{6}y + 2 = \dfrac{7}{4}$

$12\left[\dfrac{5}{6}y + 2 = \dfrac{7}{4}\right]$ which can be written as: $12\left[\dfrac{5}{6}y + 2\right] = 12\left[\dfrac{7}{4}\right]$

$12\left(\dfrac{5}{6}y\right) + 12(2) = 12\left(\dfrac{7}{4}\right)$

$\left(\dfrac{12}{1}\right)\left(\dfrac{5}{6}y\right) + 12(2) = \left(\dfrac{12}{1}\right)\left(\dfrac{7}{4}\right)$

$\left(\dfrac{60}{6}y\right) + 24 = \left(\dfrac{84}{4}\right)$

$10y + 24 = 21$

The next step is to solve for y.

$10y + 24 = 21$
$-24\ \ -24$
$10y = -3$
$\div 10\ \ \div 10$
$y = -\dfrac{3}{10}$

The solution is: $y = -\dfrac{3}{10}$

Note: You can check your solution by substituting it back into your equation.

SOLVE A LINEAR EQUATION WITH FRACTION COEFFICIENT(S)

1. Solve for x: $\dfrac{4}{5}x - 8 = 3$

   a. $x = \dfrac{23}{4}$ b. $x = \dfrac{55}{4}$ c. $x = -\dfrac{25}{4}$ d. $x = \dfrac{7}{4}$

2. Solve for t: $-\dfrac{7}{8}t - 3 = 7$

   a. $t = -\dfrac{59}{7}$ b. $t = -\dfrac{80}{7}$ c. $t = -\dfrac{32}{7}$ d. $t = -\dfrac{53}{7}$

3. Solve for r: $-\dfrac{5}{4}r + 2 = \dfrac{7}{8}$

   a. $r = \dfrac{45}{32}$ b. $r = -\dfrac{1}{2}$ c. $r = \dfrac{9}{10}$ d. $r = -\dfrac{23}{10}$

4. Solve for r: $\dfrac{7}{2}r + 3 = \dfrac{3}{8}$

   a. $r = -\dfrac{147}{16}$ b. $r = -\dfrac{3}{4}$ c. $r = 0$ d. $r = \dfrac{27}{28}$

SOLUTIONS:

1. b   2. b   3. c   4. b

SOLVE A LITERAL EQUATION

8. Solve for t: $x = -8z + 7t$

   A. $t = \frac{1}{7}x - 8z$ B. $t = \frac{1}{7}x + \frac{8}{7}z$ C. $t = \frac{1}{7}x + 8z$ D. $t = \frac{1}{7}x - \frac{8}{7}z$

SOLUTION:

Solve for t: $x = -8z + 7t$

The first step to solve for t is to move the $-8z$:

$x = -8z + 7t$
$+8z +8z$
$x + 8z = 7t$

The next step is isolate t by dividing by 7:

$(x + 8z) = 7t$
$\div 7 \div 7$
$t = \dfrac{x + 8z}{7}$

Since this solution is not one of the choices, we can simplify further by dividing each part of the numerator by 7.

$t = \dfrac{x + 8z}{7}$

$t = \dfrac{x}{7} + \dfrac{8z}{7}$

$t = \dfrac{1}{7}x + \dfrac{8}{7}z$

The solution is: $t = \dfrac{1}{7}x + \dfrac{8}{7}z$

SOLVE A LITERAL EQUATION

1. Solve $u = -3yx$ for x:

   a. $x = -\dfrac{u}{3y}$　　b. $x = \dfrac{u+3}{y}$　　c. $x = \dfrac{u}{3y}$　　d. $x = -\dfrac{1}{3}u + \dfrac{1}{3}y$

2. Solve $P = 2L + 2W$ for W:

   a. $W = 2P - L$　　b. $W = \dfrac{1}{2}P - L$　　c. $W = \dfrac{1}{2}P + L$　　d. $W = P - L$

3. Solve $x = 2v - 3y$ for y:

   a. $y = -\dfrac{1}{3}x - 2v$　　b. $y = -\dfrac{1}{3}x + 2v$　　c. $y = -\dfrac{1}{3}x - \dfrac{2}{3}v$　　d. $y = -\dfrac{1}{3}x + \dfrac{2}{3}v$

4. Solve $u = 3y - 6v$ for v:

   a. $v = -\dfrac{1}{6}u - \dfrac{1}{2}y$　　b. $v = -\dfrac{1}{6}u + 3y$　　c. $v = -\dfrac{1}{6}u + \dfrac{1}{2}y$　　d. $v = -\dfrac{1}{6}u - 3y$

Solutions:

1. a　　2. b　　3. d　　4. c

TRANSLATE A WORD PROBLEM INTO AN ALGEBRAIC EXPRESSION

9. The sum of a number and 16 is 4 less than twice the number.

   Find the equation that can be used to find this number, x.

   A. $x+16=x^2-4$ B. $x+16=4-2x$ C. $x+16=2(x-4)$ D. $x+16=2x-4$

Solution:

The sum of a number and 16 is 4 more than twice the number.

Translating each part:

The sum of a number and 16 means: $x + 16$

The word "is" means "="

Then 4 less than twice the number means: $2x - 4$

The solution is: $x + 16 = 2x - 4$

PRACTICE PROBLEMS

TRANSLATE A WORD PROBLEM INTO AN ALGEBRAIC EXPRESSION

1. If 9 times a number is increased by 20, the result is 22 less than the square of the number. Choose the equation that can be used to find this number x.

    a. $9(x+20) = x^2 - 22$
    b. $29x = x^2 - 22$
    c. $9x + 20 = 22 - x^2$
    d. $9x + 20 = x^2 - 22$

2. The sum of a number and 2 is 3 more than twice the number.
   Find the equation that can be used to find this number, x.

    a. $2x = 2x + 3$
    b. $x + 2 = 2(x+3)$
    c. $x + 2 = x^2 + 3$
    d. $x + 2 = 2x + 3$

3. If 3 times the sum of a number and 5 is equal to 7.
   Find the equation that can be used to find this number, x.

    a. $3(x+5) = 7$
    b. $3x + 5 = 7$
    c. $3x = 5 + 7$
    d. $3x + 5x = 7$

4. If 6 times a number is decreased by 8, the result is 5 less than twice the number. Choose the equation that can be used to find this number x.

    a. $6(x-8) = x^2 - 5$
    b. $6x - 8 = x^2 - 5$
    c. $6x - 8 = 2x - 5$
    d. $6x - 8 = 5 - 2x$

SOLUTIONS:

1. d
2. d
3. a
4. c

SOLVE A WORD PROBLEM

10. The length of a rectangle is 2 feet more than the width. The perimeter of the rectangle is 72 feet. Find the length.

   A. 19 feet   B. 17 feet   C. 37 feet   D. 35 feet

SOLUTION:

The first step is to translate the statements into algebraic expressions.

The length of a rectangle is 2 feet more than the width is represented by: $x + 2$

The width is represented by: x

Here is a visual representation:

```
           x + 2
      ┌─────────────┐
   x  │             │  x
      └─────────────┘
           x + 2
```

The next step is to create an algebraic equation to solve for x.

Since the perimeter is the sum of the sides, the equation is:

$(x) + (x + 2) + (x) + (x + 2) = 72$

The next step is to combine like terms:

$4x + 4 = 72$

Then solve for x:

$x = 17$

Since x is the width, then the width is 17. The length is x + 2, so the length is 19.

The answer is 19.

Note: You can check your solution by substituting it back into your algebraic equation.

SOLVE A WORD PROBLEM

1. A CD is priced at $15.00, but it is on sale for 20% off. What is the sale price of the CD?
   a. $3.00
   b. $10.00
   c. $18.00
   d. $12.00

2. If a sony play station costs $250 after a 15% discount, what was the original cost?
   a. $294.12
   b. $212.50
   c. $287.50
   d. $399.46

3. If a palm pilot costs $1300 after a 20% increase in price, what was the original cost?
   a. $1625.00
   b. $1083.33
   c. $1560.00
   d. $1040.00

4. Find the simple interest percent if you invested $1000.00 for 5 years and you received $500.00 in interest.
   a. 20%
   b. 50%
   c. 10%
   d. 40%

PRACTICE PROBLEMS

SOLUTIONS EXPLAINED IN FOLLOWING PAGES.

5. The width of a rectangular garden is 8 meters less than its length. Its perimeter is 76 meters. Find the length of the garden.

    a. 23 meters   b. 76 meters   c. 345 meters   d. 15 meters

6. The perimeter of a triangle is 51 inches. The length of the middle side is 5 inches more than the length of the smaller side and the largest side is 4 inches less than three times the length of the smallest side. Find the length of the middle side.

    a. 10 inches   b. 15 inches   c. 26 inches   d. 5 inches

7. Two shrimp boats start from the same port at the same time, but they head to opposite directions. The slower boat travels 15 knots per hour slower than the fast boat. At the end of 12 hours, they were 600 nautical miles apart. How many nautical miles had the slow boat traveled by the end of the 12–hour period?

    a. 210 nautical miles   b. 17.5 nautical miles
    c. 2.5 nautical miles   c. 390 nautical miles

The solutions explained in following pages.

PRACTICE PROBLEMS

SOLUTIONS TO PRACTICE WORD PROBLEMS

1. D

To find the discount you multiply the original price by the percent of the discount.

So, the discount is $15.00(0.20) = $3.00.

To find the sale price you subtract the original price from the discount.

So, to find the sale price you take 15.00 − 3.00 = 12.00. The sale price is $12.00.

---

2. A

Original Cost − Discount = Sale Price

$X − .15X = 250$

$1X − .15X = 250$

$\dfrac{.85X}{.85} = \dfrac{250}{.85}$

$X = 294.12$

Check

$294.12 − .15(294.12) = 250$

$294.12 − 44.12 = 250$

$250 = 250$

---

3. B

Original Cost + Increase = New Price   Check: $1083.33 + .20(1083.33) = 1300$

$X + .20X = 1300$

$1X + .20X = 1300$

$\dfrac{1.20X}{1.20} = \dfrac{1300}{1.20}$

$X = \$1083.33$

$1083.33 + 216.67 = 1300$

$1300 = 1300$

PRACTICE PROBLEMS 317

4. C

The formula to find simple interest is: Interest = Principal x Rate x Time (I = P*R*T).

The information gives would lead to: 500 = (1000)X (5).

Then simplify: 500 = X(5000).

Then solve for x by dividing both sides by 5000: 500/5000 = X(5000)/5000.

X = 0.10, the simple interest percent is 10%.

Check: 500 = 1000(0.10)(5)
         500 = 500

---

5. A

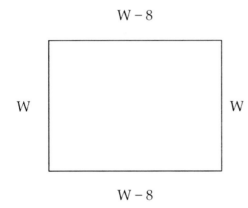

Width: W −8
Length: W
Perimeter: 76

Equation: W + W − 8 + W + W − 8 = 76

Combine Like Terms: 4W − 16 = 76
                    + 16 +16
                    4W = 92
                     4   4

Solve: W = 23

Width: 23 − 8 = 15

Length: 23

Check: 15 + 23 + 15 + 23 = 76
                          76 = 76

6. B

Small: X
Middle: X + 5
Large: 3X − 4
Perimeter: 51

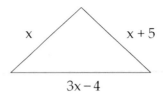

The equation is: X + X + 5 + 3X − 4 = 51

Combine like terms: 5X + 1 = 51
Solve:
$$\begin{array}{r} -1 \quad -1 \\ 5X = 50 \\ \overline{5} \quad \overline{5} \\ X = 10 \end{array}$$

Check: 10 + 10 + 5 + 3(10) − 4 = 51
10 + 15 + 26 = 51
25 + 26 = 51
51 = 51

Small: X = 10
Middle: 10 + 5 = 15
Large: 3(10) − 4 = 26

---

7. A

Type	D	R	T
Fast Boat	12X	X	12
Slow Boat	12(X−15)	X	12
Total	600		

Fill in the 4 x 4 chart with the given information.

To fill in the distance column use the formula Distance = Rate x Time

You now have the formula:

12X + 12(X − 15) = 600

Distributive Property: 12X + 12X − 180 = 600
Combine like terms: 24X − 180 = 600
$$\qquad \qquad +180 \quad +180$$
Solve for X:
$$\begin{array}{r} \dfrac{24X}{24} = \dfrac{780}{24} \\ X = 32.5 \end{array}$$

Substitute X into the chart and answer the question:

Type	D	R	T
Fast Boat	12(32.5) = 390 miles	32.5 knots	12
Slow Boat	12(32.5−15) = 210 miles	32.5−15 = 17.5 knots	12
Total	600 miles		

The answer from the chart that the slow boat traveled 210 miles.

TRANSLATE A WORD PROBLEM INTO A PROPORTION

11. Identify the proportion listed below that solves this problem:
    A car can travel 500 miles on 5 gallons of gasoline.
    How far can the car travel on 32 gallons of gasoline?

    A. $\dfrac{5}{500} = \dfrac{x}{32}$ B. $\dfrac{5}{500} = \dfrac{32}{x}$ C. $\dfrac{5}{32} = \dfrac{x}{500}$ D. $\dfrac{500}{5} = \dfrac{32}{x}$

SOLUTION:

Identify the proportion listed below that solves this problem:
A car can travel 500 miles on 5 gallons of gasoline.
How far can the car travel on 32 gallons of gasoline?

The first step is to determine how to step up the way the proportion will be written.

One way is: $\dfrac{miles}{gallons}$

So the proportions are:

$\dfrac{miles}{gallons} = \dfrac{500}{5} = \dfrac{x}{32}$

$\dfrac{500}{5} = \dfrac{x}{32}$

Since this is not an option, another way to set up the proportion is:

$\dfrac{gallons}{miles} = \dfrac{5}{500} = \dfrac{32}{x}$

The solution is: $\dfrac{5}{500} = \dfrac{32}{x}$

Remember to solve a proportion, multiply the diagonal values.

The first step to solve: $\dfrac{500}{5} = \dfrac{x}{32}$ is $500(32) = 5x$

This is the same as: $\dfrac{5}{500} = \dfrac{32}{x}$ is $500(32) = 5x$

PRACTICE PROBLEMS

TRANSLATE A WORD PROBLEM INTO A PROPORTION

1. Identify the proportion listed below that solves this problem:

    A car can travel 415 miles on 8 gallons of gasoline. How far can the car travel on 25 gallons?

    a. $\dfrac{415}{8} = \dfrac{x}{25}$
    b. $\dfrac{415}{25} = \dfrac{8}{x}$
    c. $\dfrac{8}{415} = \dfrac{x}{25}$
    d. $\dfrac{415}{x} = \dfrac{25}{8}$

2. Identify the proportion listed below that solves this problem:

    If 46 pounds of jelly beans cost 65 cents, how many pounds of jelly beans can be purchased for 130 cents?

    a. $\dfrac{65}{46} = \dfrac{x}{130}$
    b. $\dfrac{65}{46} = \dfrac{130}{x}$
    c. $\dfrac{46}{130} = \dfrac{65}{x}$
    d. $\dfrac{46}{x} = \dfrac{130}{65}$

3. Identify the proportion listed below that solves this problem:

    If Jim can eat 10 pies in 20 minutes, how many pies could Jim eat in 45 minutes?

    a. $\dfrac{45}{20} = \dfrac{x}{10}$
    b. $\dfrac{10}{20} = \dfrac{45}{x}$
    c. $\dfrac{20}{10} = \dfrac{x}{45}$
    d. $\dfrac{x}{20} = \dfrac{45}{10}$

4. Identify the proportion listed below that solves this problem:

    A shoe factory can produce 1000 pairs of shoes every 3 hours. How long would it take the shoe factory to produce 5000 pairs of shoes?

    a. $\dfrac{3}{1000} = \dfrac{5000}{x}$
    b. $\dfrac{1000}{3} = \dfrac{x}{5000}$
    c. $\dfrac{x}{3} = \dfrac{1000}{5000}$
    d. $\dfrac{1000}{3} = \dfrac{5000}{x}$

Solutions:

1. a   2. b   3. a   4. d

SIMPLIFY EXPONENTIAL EXPRESSIONS (POSITIVE INTEGER EXPONENTS)

12. Simplify: $\left(a^2b^4\right)^3\left(a^3b^4\right)$

A. $a^{10}b^{12}$ B. a^9b^{16} C. a^6b^{16} D. a^6b^{12}

SOLUTION:

Simplify: $\left(a^2b^4\right)^3\left(a^3b^4\right)$

Recall the exponent rules:
$x^2 x^4 = x^{2+4} = x^6$
$\left(x^3\right)^5 = x^{3*5} = x^{15}$
$\dfrac{x^7}{x^3} = x^{7-3} = x^4$

The first step using order of operations is to complete the exponents using exponent rules.

$\left(a^2b^4\right)^3\left(a^3b^4\right)$
$\left(a^2\right)^3\left(b^4\right)^3\left(a^3b^4\right)$
$a^6b^{12}\left(a^3b^4\right)$

The next step is to combine like terms using exponent rules.

$a^6b^{12}\left(a^3b^4\right)$
$a^6a^3b^{12}b^4$
a^9b^{16}

The solution is: a^9b^{16}

PRACTICE PROBLEMS

SIMPLIFY EXPONENTIAL EXPRESSIONS (POSITIVE INTEGER EXPONENTS)

1. Simplify: $\left(a^2 b^4\right)^2 \left(a^8 b^5\right)$

    a. $a^9 b^8$ b. $a^4 b^8$ c. $a^{12} b^{13}$ d. $a^4 b^{13}$

2. Simplify: $\left(x^4 y^3\right)^2 \left(x^5 y\right)^4$

    a. $x^{15} y^9$ b. $x^{17} y^{10}$ c. $x^{28} y^{10}$ d. $x^{28} y^{24}$

3. Simplify: $\dfrac{\left(2x^3 y^3\right)^3}{x^4}$

    a. $2x^5 y^9$ b. $8x^5 y^9$ c. $2x^{13} y^9$ d. $8x^{13} y^9$

4. Simplify: $\dfrac{\left(4x^4 y^6\right)^2}{2\left(x^2 y\right)^4}$

    a. $8y^8$ b. y^8 c. $8x^{16} y^{16}$ d. $x^{16} y^8$

Solutions:

1. c   2. c   3. b   4. a

SIMPLIFY EXPONENTIAL EXPRESSIONS (POSITIVE AND NEGATIVE INTEGER EXP.)

13. Simplify: $\dfrac{a^{-5}b^{-3}}{a^5 b^5}$

A. $\dfrac{a^{10}}{b^8}$ B. $\dfrac{1}{a^{10}b^8}$ C. $a^{10}b^8$ D. $\dfrac{b^8}{a^{10}}$

SOLUTION:

Simplify: $\dfrac{a^{-5}b^{-3}}{a^5 b^5}$

Recall the exponent rules:

$x^2 x^4 = x^{2+4} = x^6$

$\left(x^3\right)^5 = x^{3*5} = x^5$

$\dfrac{x^7}{x^3} = x^{7-3} = x^4$

$x^{-5} = \dfrac{1}{x^5}$

$\dfrac{1}{x^{-3}} = x^3$

One way to simplify is to make all of the negative exponents positive by moving them from numerator to denominator or vice versa.

$\dfrac{a^{-5}b^{-3}}{a^5 b^5}$

$\dfrac{1}{a^5 b^3 a^5 b^5}$

The next step is the combine like terms using the exponent rules.

$\dfrac{1}{a^{10} b^8}$

The solution is: $\dfrac{1}{a^{10} b^8}$

PRACTICE PROBLEMS

SIMPLIFY EXPONENTIAL EXPRESSIONS (POSITIVE AND NEGATIVE INTEGER EXP.)

1. Simplify: $\dfrac{a^{-5}b^{-3}}{a^4 b^4}$

   a. $\dfrac{1}{a^9 b^7}$ 　　b. $\dfrac{b^7}{a^9}$ 　　c. $\dfrac{a^9}{b^7}$ 　　d. $a^9 b^7$

2. Simplify: $\dfrac{x^{-5} y^{-2}}{x^5 y^{-6}}$

   a. $\dfrac{1}{x^{10} y^8}$ 　　b. y^4 　　c. $\dfrac{y^4}{x^{10}}$ 　　d. $x^{10} y^8$

3. Simplify: $\dfrac{2^{-3} x^{-1} y^3}{xy^{-4}}$

   a. $\dfrac{y^7}{8x^2}$ 　　b. $\dfrac{8y^7}{x^2}$ 　　c. $\dfrac{x^2 y^7}{8}$ 　　d. $8x^2 y^7$

4. Simplify: $\left(x^3 y^{-4}\right)^{-2}$

   a. $x^6 y^8$ 　　b. $\dfrac{y^8}{x^6}$ 　　c. $\dfrac{x}{y^6}$ 　　d. $\dfrac{1}{x^6 y^8}$

Solutions:

1. a　　2. c　　3. a　　4. b

SIMPLIFY EXPONENTIAL EXPRESSIONS (POSITIVE, NEG., AND ZERO INTEGER EXP.)

14. Simplify: $\left(a^2 b^{-4} c^0\right)^{-3}$

A. $a^{-1} b^{-7} c^{-3}$ B. $\dfrac{b^{12}}{a^6 c^3}$ C. $\dfrac{b^{12}}{a^6}$ D. $\dfrac{a^6}{b^{12}}$

Solution:

Simplify: $\left(a^2 b^{-4} c^0\right)^{-3}$

Recall the exponent rules.

$x^2 x^4 = x^{2+4} = x^6$

$\left(x^3\right)^5 = x^{3*5} = x^{15}$

$\dfrac{x^7}{x^3} = x^{7-3} = x^4$

$x^{-5} = \dfrac{1}{x^5}$

$\dfrac{1}{x^{-3}} = x^3$

$x^0 = 1$

The first step is to simplify the 0 exponent.

$\left(a^2 b^{-4} c^0\right)^{-3}$

$\left(a^2 b^{-4} (1)\right)^{-3}$

$\left(a^2 b^{-4}\right)^{-3}$

The next step is to simplify using the exponent rules.

$\left(a^2 b^{-4}\right)^{-3}$

$\left(a^2\right)^{-3} \left(b^{-4}\right)^{-3}$

$a^{-6} b^{12}$

$\dfrac{b^{12}}{a^6}$

The answer is: $\dfrac{b^{12}}{a^6}$

SIMPLIFY EXPONENTIAL EXPRESSIONS (POSITIVE, NEG., AND ZERO INTEGER EXP.)

1. Simplify: $(y^3 x^0)^{-8}$

   a. y^{24}
   b. $y^{11} x^8$
   c. $\dfrac{1}{y^{24}}$
   d. $\dfrac{1}{y^5 x^8}$

2. Simplify: $(u^5 y^{-2} z^0)^{-3}$

   a. $\dfrac{y^6}{u^{15}}$
   b. $\dfrac{y^6}{u^{15} z^3}$
   c. $\dfrac{u^2}{y^5 z^3}$
   d. $\dfrac{1}{u^{15} y^6}$

3. Simplify: $(2a^3 b^0)^{-4}$

   a. $\dfrac{a^{12}}{8b^4}$
   b. $8a^{12} b^4$
   c. $\dfrac{-16}{a^{12}}$
   d. $\dfrac{1}{16a^{12}}$

4. Simplify: $3x^2 (x^{-5} y^0 z^2)^{-4}$

   a. $3x^{22} y^4 z^8$
   b. $\dfrac{3x^{22}}{y^4 z^8}$
   c. $\dfrac{3x^{22}}{z^8}$
   d. $\dfrac{3}{x^{18} z^8}$

Solutions:

1. c   2. a   3. d   4. c

SCIENTIFIC NOTATION (TO AND FROM)

15. Convert to scientific notation: 0.000000195

A. 0.195×10^{-6}　　B. 1.95×10^{-7}　　C. 1.95×10^{7}　　D. 1.95×10^{-6}

Solution:

Convert to scientific notation: 0.000000195

The first step is to determine where the decimal place should be, to convert this number from standard notation to scientific notation. In scientific notation the decimal should be after the first non zero from the left.

In this case the decimal should be after the 1, which is 7 movements to the right.

If the decimal moved to the right, the exponent is negative, to the left the exponent is positive.

Then put in this form: 1.95×10^{-7}

The solution is: 1.95×10^{-7}

Note: This can be checked by converting from scientific notation to standard notation.

To convert 1.95×10^{-7} to standard notation, start at the decimal point; the −7 means move the decimal point 7 movements to the left (due to the negative). A positive exponent means move the decimal to the right, a negative means move to the left.

1.95×10^{-7}
0.000000195

PRACTICE PROBLEMS

SCIENTIFIC NOTATION (TO AND FROM)

1. Convert to standard form: 5.19×10^7

    a. 0.000000519   b. 51900000   c. 0.0000000519   d. 5190000000

2. Convert to standard form: -6.4×10^{-4}

    a. 0.00064   b. −64000   c. 64000   d. −0.00064

3. Convert to scientific notation: −1640000

    a. -1.64×10^6 b. 0.164×10^7 c. -1.64×10^{-6} d. 1.64×10^7

4. Convert to scientific notation: 0.000231

    a. 2.31×10^{-4} b. 2.31×10^{-3} c. 0.231×10^{-3} d. 2.31×10^4

Solutions:

1. b   2. d   3. a   4. a

ADDITION/SUBTRACTION OF POLYNOMIALS

16. Simplify: $(7x^2 - 8x + 4) - (6x^2 - 8x - 5)$

    A. $x^2 - 16x - 1$ B. $x^2 - 1$ C. $x^2 + 1$ D. $x^2 + 9$

Solution:

Simplify: $(7x^2 - 8x + 4) - (6x^2 - 8x - 5)$

One way to simplify is to rewrite this problem by distributing the subtraction sign:
$(7x^2 - 8x + 4) - (6x^2 - 8x - 5)$
$7x^2 - 8x + 4 - 6x^2 - (-8x) - (-5)$
$7x^2 - 8x + 4 - 6x^2 + 8x + 5$

The next step is to combine like terms.
$7x^2 - 8x + 4 - 6x^2 + 8x + 5$
$7x^2 - 6x^2 - 8x + 8x + 4 + 5$
$x^2 + 9$

The solution is: $x^2 + 9$

PRACTICE PROBLEMS

ADDITION/SUBTRACTION OF POLYNOMIALS

1. Simplify: $(9x^2 - 8x - 2) + (4x^2 - 3x + 3)$

   a. $13x^2 - 11x + 1$ b. $13x^2 + 5x + 1$ c. $13x^2 - 11x + 5$ d. $5x^2 - 5x - 5$

2. Simplify: $(7x^2 + 4x - 7) - (2x^2 - 6x + 5)$

   a. $5x^2 + 10x - 2$ b. $5x^2 - 2x - 12$ c. $5x^2 - 2x - 2$ d. $5x^2 + 10x - 12$

3. Simplify: $(3x^2 - 5x - 6) + (-x + 7)$

   a. $3x^2 - 4x - 13$ b. $3x^2 - 6x - 13$ c. $3x^2 - 6x + 1$ d. $3x^2 - 4x + 1$

4. Simplify: $(x^2 - 5) - (2x^2 - 7x + 5)$

   a. $-x^2 + 7x - 10$ b. $-x^2 - 7x - 10$ c. $-x^2 - 7x$ d. $x^2 + 7x$

Solutions:

1. a 2. d 3. c 4. a

MULTIPLY A MONOMIAL AND A BINOMIAL

17. Simplify: $-7x(-4x+7)$

A. $28x^2 - 49x$ B. $28x^2 + 49x$ C. $-28x^2 - 49x$ D. $-21x^2$

SOLUTION:

Simplify: $-7x(-4x+7)$

The first step to simplify is to use the distributive property and simplify.

$-7x(-4x+7)$
$-7x(-4x)+(-7x)(7)$
$28x^2 - 49x$

The solution is: $28x^2 - 49x$

MULTIPLY A MONOMIAL AND A BINOMIAL

1. Simplify: $9x(5x-2)$

   a. $27x^2$ b. $14x^2-18x$ c. $45x^2+18x$ d. $45x^2-18x$

2. Simplify: $-8x(-6x+6)$

   a. $48x^2-48x$ b. $-48x^2-48x$ c. $48x^2+48x$ d. 0

3. Simplify: $-2x^2(x^3-5x)$

   a. $-2x^6+10x^2$ b. $-2x^5-10x^3$ c. $-2x^5+10x^3$ d. $-2x^5+10x^2$

4. Simplify: $4x^3y^5(-5x^2y^3+x)$

   a. $-20x^5y^8+4x^4y^5$ b. $-20x^6y^{15}+4x^3y^5$

   c. $20x^5y^8-4x^4y^5$ d. $-20x^6y^{15}+4x^3$

Solutions:

1. d   2. a   3. c   4. a

MULTIPLY TWO BINOMIALS

18. Simplify: $(4x-7)(6x-7)$

   A. $24x^2 + 14x + 49$ B. $24x^2 - 70x + 49$ C. $24x^2 - 14x - 49$ D. $24x^2 - 70x - 49$

Solution:

Simplify: $(4x-7)(6x-7)$

One way to simplify this problem is to use the FOIL method:

First: $(4x)(6x)$: $24x^2$

Outer: $(4x)(-7)$: $-28x$

Inner: $(-7)(6x)$: $-42x$

Last: $(-7)(-7)$: 49

The result is: $24x^2 - 28x - 42x + 49$

Then combine like terms: $-28x - 42x = -70x$

The solution is: $24x^2 - 70x + 49$

PRACTICE PROBLEMS

MULTIPLY TWO BINOMIALS

1. Simplify: $(2x-3)(6x-3)$

   a. $12x^2 - 24x - 9$ b. $12x^2 + 12x + 9$ c. $12x^2 - 24x + 9$ d. $12x^2 - 12x - 9$

2. Simplify: $(9x+6)(-9x-2)$

   a. $-81x^2 - 72x - 12$
   c. $-81x^2 - 36x - 12$
   b. $-81x^2 + 72x - 12$
   d. $81x^2 - 72x - 12$

3. Simplify: $(-x-3)(-2x-5)$

   a. $-2x^2 - 11x - 15$ b. $2x^2 - 11x + 15$ c. $2x^2 + 1x - 15$ d. $2x^2 + 11x + 15$

4. Simplify: $(3x+8)(7x^2+6x-3)$

   a. $21x^3 + 74x^2 + 39x - 24$
   c. $21x^3 + 74x^2 + 57x + 24$
   b. $21x^3 - 74x^2 + 57x - 24$
   d. $21x^3 - 38x^2 + 57x - 24$

SOLUTIONS:

1. c   2. a   3. d   4. a

FACTORING A POLYNOMIAL: GREATEST COMMON FACTOR (GCF)

19. Factor completely: $30y^{50}s^8 + 14y^{55}s^6 + 24y^{15}s^6$

A. $2y^5s^2\left(15y^{10}s^4 + 7y^{11}s^3 + 12y^3s^3\right)$ B. $2ys\left(15y^{49}s^7 + 7y^{54}s^5 + 12y^{14}s^5\right)$

C. $2y^{14}s^5\left(15y^{36}s^3 + 7y^{41}s^1 + 12y^1s^1\right)$ D. $2y^{15}s^6\left(15y^{35}s^2 + 7y^{40} + 12\right)$

SOLUTION:

Factor completely: $30y^{50}s^8 + 14y^{55}s^6 + 24y^{15}s^6$

The first step to factor is to identify which method should be used. In this case the factoring method is by pulling out the Greatest Common Factor.

To determine the GCF of the coefficients we look at the prime factors of the coefficients:

$30y^{50}s^8 + 14y^{55}s^6 + 24y^{15}s^6$

$2\cdot3\cdot5y^{50}s^8 + 2\cdot7y^{55}s^6 + 2\cdot2\cdot2\cdot3y^{15}s^6$

The Greatest Common Factor of the coefficients is: 2

To determine the Greatest Common Factor of the variables, we look for the most of each variable in each term. All of the terms must have the variable for it to be factored out. It will be the smallest exponent as long as all terms have the variable.

$30y^{50}s^8 + 14y^{55}s^6 + 24y^{15}s^6$

The Greatest Common Factor for the variable y is: y^{15}

The Greatest Common Factor for the variable s is: s^6

The Greatest Common Factor is: $2y^{15}s^6$

If we factor out $2y^{15}s^6$ from each term, we need to consider what is remaining.

If we divide each term by $2y^{15}s^6$, that will give the remaining.

$$\frac{30y^{50}s^8}{2y^{15}s^6} + \frac{14y^{55}s^6}{2y^{15}s^6} + \frac{24y^{15}s^6}{2y^{15}s^6}$$

$15y^{35}s^2 \ + \ 7y^{40} \ + \ 12$

The solution is: $2y^{15}s^6\left(15y^{35}s^2 + 7y^{40} + 12\right)$

Note: This solution can be checked by using the distributive property but be sure to pull out the greatest common factor.

FACTORING A POLYNOMIAL: GREATEST COMMON FACTOR (GCF)

1. Factor: $16x^{20}s^{12} - 16x^{25}s^4$

   a. $16x^5s^4\left(x^4s^3 - x^5s\right)$
   b. $16x^{20}s^4\left(s^8 - x^5\right)$
   c. $16x^{20}s^5\left(s^8 - 2x^5s\right)$
   d. $16x^{21}s^4\left(xs^8 - x^5s\right)$

2. Factor: $24z^{15}t^9 + 24z^{10}t^3$

   a. $24z^5t^3\left(z^3t^3 + z^2t\right)$
   b. $24z^{10}t^3\left(z^5t^6 + 1\right)$
   c. $24z^{16}t^3\left(zt^6 + t\right)$
   d. $24z^{10}t^4\left(z^5t^6 + 3t\right)$

3. Factor: $36y^{30}v^8 + 20y^{15}v^{16} + 48y^{21}v^4$

   a. $4yv\left(9y^{29}v^7 + 5y^{14}v^{15} + 12y^{20}v^3\right)$
   b. $4y^3v^2\left(9y^{10}v^4 + 5y^5v^8 + 12y^7v^2\right)$
   c. $4y^{14}v^3\left(9y^{16}v^5 + 5yv^{13} + 12y^7v\right)$
   d. $4y^{15}v^4\left(9y^{15}v^4 + 5v^{12} + 12y^6\right)$

4. Factor: $55z^{24}v^{45} + 70z^{36}v^{40} + 30z^{32}v^{55}$

   a. $5zv\left(11z^{23}v^{44} + 14z^{35}v^{39} + 6z^{31}v^{54}\right)$
   b. $5z^{24}v^{40}\left(11v^5 + 14z^{12} + 6z^8v^{15}\right)$
   c. $5z^{23}v^{39}\left(11zv^6 + 14z^{13}v + 6z^9v^{16}\right)$
   d. $5z^4v^5\left(11z^6v^9 + 14z^9v^8 + 6z^8v^{11}\right)$

SOLUTIONS:

1. b   2. b   3. d   4. b

PRACTICE PROBLEMS

FACTORING A POLYNOMIAL: DIFFERENCE OF TWO SQUARES

20. Factor completely: $25z^2 - 64x^2$

A. $(5z+16x)(5z-4x)$ B. $(5z-8x)^2$

C. $(5z+8x)(5z-8x)$ D. $(5z+4x)(5z-16x)$

SOLUTION:

Factor completely: $25z^2 - 64x^2$.

The first step to factor is to identify which method should be used. In this case the factoring method is the difference of two squares. It's because there are only 2 terms, there is a subtraction sign in the middle, and the front and back terms can be square rooted.

Make sure the expression can be factored by pulling out the Greatest Common Factor first. If you can factor by pulling out the GCF, that should be completed first.

The first step is to take the square root of the front term is:

$25z^2$

$\sqrt{25z^2}$

$5z$

Then take the square root of the back term is:

$64x^2$

$\sqrt{64x^2}$

$8x$

The solution is: $(5z+8x)(5z-8x)$

Note: This solution can be checked by simplifying through FOIL.

$(5z+8x)(5z-8x)$
$5z(5z)+(5z)(-8x)+(8x)(5z)+(8x)(-8x)$
$25z^2 - 40xz + 40xz - 64x^2$
$25z^2 - 64x^2$

FACTORING A POLYNOMIAL: DIFFERENCE OF TWO SQUARES

1. Factor: $t^2 - 36x^2$

   a. $(t-6x)(t+6x)$ b. $(t+3x)(t-12x)$ c. $(t+12x)(t-3x)$ d. $(t-6x)^2$

2. Factor: $25y^2 - 81$

   a. $(5y-9)(5y-9)$ b. $(5y+9)(5y+9)$ c. $(25y-9)(y+9)$ d. $(5y-9)(5y+9)$

3. Factor: $x^2 - y^2$

   a. $(x-y)(x-y)$ b. $(1-y^2)(x^2+1)$ c. $(x-y)(x+y)$ d. $(x+y)(x+y)$

4. Factor: $50x^2 - 32y^2$

   a. $(25x+16y)(25x-16y)$ b. $2(5x+4y)(5x-4y)$
   c. $2(5x-4y)(5x-4y)$ d. $(25x-16y)(25x-16y)$

Solutions:

1. a   2. d   3. c   4. b

PRACTICE PROBLEMS

FACTORING A POLYNOMIAL: BY GROUPING

21. Factor completely: $yr + yk + br + bk$

   A. $(y-b)(r-k)$ B. $(y+b)(r+k)$

   C. $(y+b)(r-k)$ D. $(5z+4x)(5z-16x)$

SOLUTION:

Factor completely: $yr + yk + br + bk$

The first step to factor is to identify which method should be used. In this case the factoring method is grouping because there are 4 terms, the Greatest Common Factor can not be factored out of all 4 terms, but we can group terms and then factor out a Greatest Common Factor.

The next step is to group terms that have similar factors.

$yr + yk + br + bk$
$yr + yk \quad + br + bk$

The next step is to factor out similar terms from each group.

$yr + yk \qquad\qquad +br + bk$
$y(r+k) \qquad\qquad +b(r+k)$

The next step is to notice that both groups now have a common factor of $(r + k)$.

Factor out the $(r + k)$ from both group and show what remains.

$y(r+k) \quad +b(r+k)$
$(r+k)(y+b)$

The solution is: $(y+b)(r+k)$

Note: This solution can be checked by simplifying through FOIL.

$(y+b)(r+k)$
$y(r) + (y)(k) + (b)(r) + (b)(k)$
$yr + yk + br + bk$

FACTORING A POLYNOMIAL: BY GROUPING

1. Factor: $zs - zf + as - af$

   a. $(s-f)(z+a)$ b. $(s-f)(z-a)$ c. $(z-s)(f+a)$ d. $(z+a)(s+f)$

2. Factor: $ws - wh - cs + ch$

   a. $(w+c)(s-h)$ b. $(w+c)(s+h)$ c. $(w-c)(s-h)$ d. $(w-c)(s+h)$

3. Factor: $9y^2 - 12ys + 3y - 4s$

   a. $(3y+s)(3y-4)$ b. $(3y-1)(3y-4s)$ c. $(3y+1)(3y-4s)$ d. $(3y+1)(3y+4s)$

4. Factor: $6z^2 - 2zs - 3z + s$

   a. $(2z-1)(3z-s)$ b. $(2z-1)(3z+s)$ c. $(6z-1)(z+s)$ d. $(2z-s)(3z+1)$

Solutions:

1. a   2. c   3. c   4. a

FACTORING A POLYNOMIAL: A TRINOMIAL

22. Identify a factor of the following trinomial: $6t^2 - 25t + 4$

A. $(3t+2)$ B. $(3t-2)$ C. $(t-4)$ D. $(t-1)$

SOLUTION:

Identify a factor of the following trinomial: $6t^2 - 25t + 4$

The first step in factoring is to identify the best method for factoring. In this case the best method is factoring a trinomial because a trinomial is given.

There are two ways to factor this polynomial: <u>Trial and Error</u> and <u>Change to Grouping</u>

First Method: Trial and Error

We know that when we factor $6t^2 - 25t + 4$, the factored form will be: ( )( )

We will first list the possible factors:

We must get the leading term of $6t^2$

The possibilities are: $(1t)(6t)$ $(2t)(3t)$

We know that it must multiply to the last term of + 4.

The possibilities with signs are: (+1)(+4)   (−1)(−4) (+2)(+2)   (−2)(−2)

Since the middle term is negative, and it must add to the middle term, then it must be either (−1)(−4) or (−2)(−2).

Then try different combinations and simply each one to try get $6t^2 - 25t + 4$:

Ex. $(2t-2)(3t-2)$ $(2t-1)(3t-4)$ $(t-2)(6t-2)$ $(6t-1)(t-4)$

The solution is: $(6t-1)(t-4)$.

Since the question only requires one of the factors, the solution is: $(t-4)$

Note: **This solution can be checked by simplifying through FOIL.**

$(6t-1)(t-4)$

$6t(t) + (6t)(-4) + (-1)(t) + (-1)(-4)$

$6t^2 - 24t - 1t + 4$

$6t^2 - 25t + 4$

Second Method: Change to grouping

SOLUTION:

Identify a factor of the following trinomial: $6t^2 - 25t + 4$.

In this method we want the trinomial to become a polynomial with four terms so we can perform grouping. There is a specific way to change the trinomial into a polynomial with four terms.

Step 1: Find the middle factors.

Multiply the leading coefficient by the last constant term.

$6 \cdot 4 = +24$, we want to find factors that multiply to get this number (including the sign).

The factors must also add to the middle term coefficient, in this case -25.

The factors that multiply to $+24$ and add to -25 are: -24 and -1

Step 2: Rewrite the polynomial.

We found the two coefficients to replace the middle term to form a polynomial with 4 terms.

The trinomial $6t^2 - 25t + 4$ now becomes: $6t^2 - 24t - 1t + 4$.

It does not matter which number goes first, but since we are going to group you want to place factors next to terms they have something in common with. For example, place the "$-24t$" next to the "$16t^2$" because they have a "$6t$" in common.

Step 3: Use Grouping

$6t^2 - 24t - 1t + 4$

$6t^2 - 24t \qquad - 1t + 4$

$6t(t - 4) \qquad -1(t - 4)$

$(t - 4)(6t - 1)$

The solution is: $(6t - 1)(t - 4)$

Since the question only requires one of the factors, the solution is: $(t - 4)$

Note: This solution can be checked by simplifying through FOIL.

$(6t - 1)(t - 4)$

$6t(t) + (6t)(-4) + (-1)(t) + (-1)(-4)$

$6t^2 - 24t - 1t + 4$

$6t^2 - 25t + 4$

FACTORING A POLYNOMIAL: A TRINOMIAL (MIXED)

1. Factor: $x^2 + 7x + 10$

   a. $(x+5)(x+2)$ b. $(x+5)(x-2)$ c. $(x+10)(x+1)$ d. $(x-5)(x-2)$

2. Factor: $x^2 + x - 42$

   a. $(x-7)(x-6)$ b. $(x+21)(x-2)$ c. $(x+6)(x-7)$ d. $(x+7)(x-6)$

3. Factor: $2y^2 - 7y + 3$

   a. $(2y-1)(y+3)$ b. $(2y-1)(y-3)$ c. $(2y+1)(y+3)$ d. $(y+3)(2y-1)$

4. Factor: $8y^2 - 2y - 15$

   a. $(4y-5)(2y+3)$ b. $(8y+5)(y-3)$ c. $(2y-3)(4y+5)$ d. $(4y-3)(2y+5)$

Solutions:

1. a
2. d
3. b
4. c

344 PRACTICE PROBLEMS

FACTORING A POLYNOMIAL: A TRINOMIAL

1. Identify a factor of the following trinomial: $5x^2 + 11x + 2$.

   a. $(5x+2)$ b. $(x-2)$ c. $(x+1)$ d. $(5x+1)$

2. Identify a factor of the following trinomial: $6r^2 - 23r - 4$.

   a. $(6r+1)$ b. $(3r-2)$ c. $(r+1)$ d. $(r-1)$

3. Identify a factor of the following trinomial: $10x^2 + 21x - 10$.

   a. $(5x+2)$ b. $(10x-1)$ c. $(2x+5)$ d. $(2x-5)$

4. Identify a factor of the following trinomial: $9x^2 - 32x + 15$.

   a. $(9x-5)$ b. $(3x-5)$ c. $(x+3)$ d. $(9x+5)$

SOLUTIONS:

1. d   2. a   3. c   4. a

SIMPLIFY A RATIONAL EXPRESSION: REDUCE BY FACTORING

23. Simplify: $\dfrac{2x^2 - 5x + 3}{x^2 - 1}$.

   A. $\dfrac{2x-3}{x+1}$
   B. $\dfrac{2x+3}{x+1}$
   C. $\dfrac{x-2}{x+1}$
   D. $\dfrac{2x-3}{2x-3}$

Solution:

Simplify: $\dfrac{2x^2 - 5x + 3}{x^2 - 1}$

The first step to simplify this expression is factor the numerator and denominator if possible.

$\dfrac{2x^2 - 5x + 3}{x^2 - 1}$

$\dfrac{(x-1)(2x-3)}{(x-1)(x+1)}$

The next step is to simplify by looking for factors that appear in both the numerator and denominator.

$\dfrac{(x-1)(2x-3)}{(x-1)(x+1)}$

$\dfrac{(2x-3)}{(x+1)}$

The solution is: $\dfrac{(2x-3)}{(x+1)}$

SIMPLIFY A RATIONAL EXPRESSION: REDUCE BY FACTORING

1. Simplify: $\dfrac{x^2+8x+7}{x^2+4x-21}$

   a. $\dfrac{x-1}{x-3}$ b. $\dfrac{x+1}{x-3}$ c. $\dfrac{x+1}{x+3}$ d. $\dfrac{x-1}{x+3}$

2. Simplify: $\dfrac{2x^2-8x}{x^2-6x+8}$

   a. $\dfrac{2x}{x-2}$ b. $\dfrac{2x}{x+2}$ c. $\dfrac{x-4}{x-2}$ d. $\dfrac{x}{x+2}$

3. Simplify: $\dfrac{3x^2-7x-6}{12x^2+5x-2}$

   a. $\dfrac{x-1}{4x-1}$ b. $\dfrac{x-3}{2x+1}$ c. $\dfrac{x-3}{4x-1}$ d. $\dfrac{x+3}{4x-1}$

4. Simplify: $\dfrac{3x^2+14x+8}{9x^2+3x-2}$

   a. $\dfrac{x+4}{3x+2}$ b. $\dfrac{x-4}{3x-1}$ c. $\dfrac{x+4}{3x-1}$ d. $\dfrac{3x+4}{3x-1}$

Solutions:

1. b   2. a   3. c   4. c

SOLVING QUADRATIC EQUATIONS BY FACTORING (LEADING COEFFICIENT IS ONE, A=1)

24. Simplify: $x^2 + 6x + 5 = 0$

A. $x = \dfrac{1}{5}, x = -1$ B. $x = -5, x = 1$ C. $x = -5, x = -1$ D. $x = 5, x = 1$

Solution:

Solve: $x^2 + 6x + 5 = 0$.

The first step is to factor the left side of the equation.

$x^2 + 6x + 5 = 0$
$(x+5)(x+1) = 0$

The next step is to set each factor equal to 0.

$(x+5)(x+1) = 0$
$x + 5 = 0 \quad x + 1 = 0$

The next step is to solve each equation.

$x + 5 = 0 \quad\quad x + 1 = 0$
$-5 \ -5 \quad\quad\quad -1 \ -1$
$x = -5 \quad\quad\quad x = -1$

The solutions are: $x = -5 \quad x = -1$

Note: The solutions can be checked by substituting them back into the original equation.

$x^2 + 6x + 5 = 0 \quad\quad\quad x^2 + 6x + 5 = 0$
$(-5)^2 + 6(-5) + 5 = 0 \quad\quad (-1)^2 + 6(-1) + 5 = 0$
$0 = 0 \quad\quad\quad\quad\quad\quad\quad 0 = 0$

SOLVING QUADRATIC EQUATIONS BY FACTORING (LEADING COEFFICIENT IS ONE, A=1)

1. Solve: $x^2 + 5x + 6 = 0$

    a. $x = 2, x = 3$ b. $x = \dfrac{1}{2}, x = -3$ c. $x = -2, x = -3$ d. $x = -2, x = 3$

2. Solve: $x^2 - 2x - 15 = 0$

    a. $x = -5, x = 3$ b. $x = -\dfrac{1}{5}, x = -3$ c. $x = 5, x = -3$ d. $x = 5, x = 3$

3. Solve: $x^2 + 2x - 24 = 0$

    a. $x = -6, x = 4$ b. $x = 6, x = 4$ c. $x = -6, x = -4$ d. $x = -12, x = 2$

4. Solve: $x^2 + 10x + 25 = 0$

    a. $x = 5, x = -5$ b. $x = 5$ c. $x = -5$ d. $x = -1, x = -25$

SOLUTIONS:

1. c   2. c   3. a   4. c

SOLVING QUADRATIC EQUATIONS BY FACTORING (LEADING COEFFICIENT IS NOT ONE)

25. Solve: $35x^2 + 11x - 6 = 0$

A. $x = -\dfrac{3}{5}, x = -\dfrac{2}{7}$ B. $x = \dfrac{5}{3}, x = \dfrac{2}{7}$ C. $x = \dfrac{3}{5}, x = -\dfrac{2}{7}$ D. $x = -\dfrac{3}{5}, x = \dfrac{2}{7}$

SOLUTION:

Solve: $35x^2 + 11x - 6 = 0$

The first step is to factor the left side of the equation.

$35x^2 + 11x - 6 = 0$
$(5x + 3)(7x - 2) = 0$

The next step is to set each factor equal to 0.
$(5x + 3)(7x - 2) = 0$
$5x + 3 = 0 \qquad 7x - 2 = 0$

The next step is to solve each equation.
$(5x + 3)(7x - 2) = 0$
$5x + 3 = 0 \qquad\quad 7x - 2 = 0$
$-3 \;-3 \qquad\quad +2 \;+2$
$5x = -3 \qquad\qquad 7x = 2$
$\div 5 \;\div 5 \qquad\qquad \div 7 \;\div 7$
$x = -\dfrac{3}{5} \qquad\qquad x = \dfrac{2}{7}$

The solutions are: $x = -\dfrac{3}{5} \qquad x = \dfrac{2}{7}$

Note: The solutions can be checked by substituting them **back into the original equation**.

$35x^2 + 11x - 6 = 0 \qquad\qquad 35x^2 + 11x - 6 = 0$

$35\left(-\dfrac{3}{5}\right)^2 + 11\left(-\dfrac{3}{5}\right) - 6 = 0 \qquad 35\left(\dfrac{2}{7}\right)^2 + 11\left(\dfrac{2}{7}\right) - 6 = 0$

$\phantom{35\left(-\dfrac{3}{5}\right)^2 + 11}0 = 0 \qquad\qquad\qquad\qquad 0 = 0$

SOLVING QUADRATIC EQUATIONS BY FACTORING (LEADING COEFFICIENT IS NOT ONE)

1. Solve: $2x^2 + x - 21 = 0$

    a. $x = -3, x = \dfrac{7}{2}$ b. $x = 3, x = \dfrac{7}{2}$ c. $x = 3, x = -\dfrac{7}{2}$ d. $x = -\dfrac{1}{3}, x = \dfrac{7}{2}$

2. Solve: $7x^2 - 13x - 2 = 0$

    a. $x = -\dfrac{1}{2}, x = -\dfrac{1}{7}$ b. $x = 2, x = -\dfrac{1}{7}$ c. $x = 2, x = \dfrac{1}{7}$ d. $x = -2, x = \dfrac{1}{7}$

3. Solve: $25x^2 + 25x - 14 = 0$

    a. $x = \dfrac{5}{7}, x = \dfrac{2}{5}$ b. $x = -\dfrac{7}{5}, x = \dfrac{2}{5}$ c. $x = -\dfrac{7}{5}, x = -\dfrac{2}{5}$ d. $x = \dfrac{7}{5}, x = -\dfrac{2}{5}$

4. Solve: $5x^2 - 28x - 49 = 0$

    a. $x = 7, x = \dfrac{5}{7}$ b. $x = -7, x = \dfrac{7}{5}$ c. $x = -7, x = -\dfrac{7}{5}$ d. $x = 7, x = -\dfrac{7}{5}$

Solutions:

1. c
2. b
3. b
4. d

SIMPLIFY SQUARE ROOT OF A MONOMIAL

26. Simplify completely: $\sqrt{18x^7w^4}$

A. $9w^2x^3\sqrt{2x^7}$ B. $3x^3w^2\sqrt{2x}$ C. $3w^2\sqrt{2x^7}$ D. $3w^2x^3\sqrt{2x}$

SOLUTION:

Simplify completely: $\sqrt{18x^7w^4}$

One method of simplifying this radical expression is to break the radicand into prime factors.

$\sqrt{18x^7w^4}$

$\sqrt{2 \cdot 3 \cdot 3xxxxxxxwwww}$

The next step is for every pair, one value is brought out the radical, then simplify.

$\sqrt{2 \cdot 3 \cdot 3\ xx\ xx\ xx\ x\ ww\ ww}$

$3xxxww\sqrt{2x}$

$3x^3w^2\sqrt{2x}$

Another method is to create two radicals where one radical contains what can be square rooted (perfect squares) and the other radical contains what can not be square rooted (non perfect squares).

$\sqrt{18x^7w^4}$

$\sqrt{9x^6w^4}\ \sqrt{2x}$

The next step is to take the square root of the perfect squares.

$\sqrt{9x^6w^4}\ \sqrt{2x}$

$3x^3w^2\sqrt{2x}$

The solution is: $3x^3w^2\sqrt{2x}$

SIMPLIFY SQUARE ROOT OF A MONOMIAL

1. Simplify completely: $\sqrt{25x^6y^5}$

   a. $5x^3y^2\sqrt{y}$ b. $25x^3y^2\sqrt{y}$ c. $5x^2y^2\sqrt{y}$ d. $5x^3y^2\sqrt{y^3}$

2. Simplify completely: $\sqrt{45x^3y^{10}z}$

   a. $9x^2y^{10}\sqrt{5xz}$ b. $3xy^5z\sqrt{5x}$ c. $3xy^5\sqrt{5xz}$ d. $9xy^5\sqrt{5xz}$

3. Simplify completely: $2\sqrt{64x^9u^2}$

   a. $128ux^4\sqrt{x}$ b. $16ux^4\sqrt{x}$ c. $16u\sqrt{x^9}$ d. $128ux^4\sqrt{x^9}$

4. Simplify completely: $5x\sqrt{72x^7y^{12}z}$

   a. $6x^3y^6\sqrt{2xz}$ b. $15x^3y^6\sqrt{8xz}$ c. $30x^4y^3z\sqrt{2x}$ d. $30x^4y^6\sqrt{2xz}$

Solutions:

1. a   2. c   3. b   4. d

SIMPLIFY SQUARE ROOT OF A POLYNOMIAL USING THE DISTRIBUTIVE PROPERTY

27. Simplify: $3\sqrt{2}\left(\sqrt{10} - 5\sqrt{2}\right)$

A. $3\sqrt{20} - 60$ B. $-24\sqrt{5}$ C. $12\sqrt{5} - 15\sqrt{4}$ D. $6\sqrt{5} - 30$

Solution:

Simplify:

$3\sqrt{2}\left(\sqrt{10} - 5\sqrt{2}\right)$

The first step to simplify this expression is to use the distributive property:

$3\sqrt{2}\left(\sqrt{10} - 5\sqrt{2}\right)$
$\left(3\sqrt{2}\right)\left(\sqrt{10}\right) + \left(3\sqrt{2}\right)\left(5\sqrt{2}\right)$

The next step is to multiply the radical expressions. Remember to multiply radical expressions, multiply the coefficients and multiply the radicands.

$\left(3\sqrt{2}\right)\left(1\sqrt{10}\right) + \left(3\sqrt{2}\right)\left(-5\sqrt{2}\right)$
$3 \cdot 1\sqrt{2 \cdot 10} + 3 \cdot -5\sqrt{2 \cdot 2}$
$3\sqrt{20} - 15\sqrt{4}$

The next step is to simply, if needed, the radical terms.

$3\sqrt{20} - 15\sqrt{4}$
$3\sqrt{4}\sqrt{5} - 15\sqrt{4}$
$3 \cdot 2\sqrt{5} - 15 \cdot 2$
$6\sqrt{5} - 30$

The solution is: $6\sqrt{5} - 30$

PRACTICE PROBLEMS

SIMPLIFY SQUARE ROOT OF A POLYNOMIAL USING THE DISTRIBUTIVE PROPERTY

1. Simplify completely: $\sqrt{5}\left(\sqrt{2}+5\sqrt{7}\right)$

   a. $\sqrt{10}+5\sqrt{7}$ b. $5+3\sqrt{5}$ c. $15\sqrt{5}$ d. $\sqrt{10}+5\sqrt{35}$

2. Simplify completely: $2\sqrt{3}\left(\sqrt{6}-4\sqrt{3}\right)$

   a. $6\sqrt{2}-24$ b. $2\sqrt{18}-72$ c. $12\sqrt{3}-8\sqrt{9}$ d. $6\sqrt{2}-72$

3. Simplify completely: $\left(2\sqrt{5}-4\right)\left(\sqrt{5}+3\right)$

   a. $2\sqrt{5}+2$ b. $-2+10\sqrt{5}$ c. $-2+2\sqrt{5}$ d. $2\sqrt{25}+10\sqrt{5}-12$

4. Simplify completely: $\left(\sqrt{6}-2\sqrt{3}\right)\left(\sqrt{6}+2\sqrt{3}\right)$

   a. -6 b. $-6-4\sqrt{18}$ c. $-6-12\sqrt{3}$ d. $6+12\sqrt{3}$

SOLUTIONS:

1. d   2. a   3. c   4. a

SOLVING A LINEAR INEQUALITY

28. Solve the inequality: $14x + 4 \leq 26x + 20$.

   A. $x \geq \dfrac{4}{3}$ B. $x \leq \dfrac{4}{3}$ C. $x \geq -\dfrac{4}{3}$ D. $x \leq -\dfrac{4}{3}$

SOLUTION:

Solve the inequality: $14x + 4 \leq 26x + 20$.

The first step is to get the variables on the same side. Since this is an inequality it is preferred to have the variables on the left side because inequalities are generally written with the variable on the left so they are easier to read.

$$14x + 4 \leq 26x + 20$$
$$-26x \quad\quad -26x$$
$$-12x + 4 \leq 20$$

The next step is to solve for x. Remember if both sides of an inequality are multiplied or divided by a negative number, then the inequality changes from less than to greater than or from greater than to less than.

$$-12x + 4 \leq 20$$
$$\quad\quad -4 \quad -4$$
$$-12x \quad\quad \leq 16$$
$$\div -12 \quad \div -12$$
$$x \geq -\dfrac{16}{12}$$
$$x \geq -\dfrac{4}{3}$$

The solution is: $x \geq -\dfrac{4}{3}$

Note: This solution can be checked if you substitute a value that is $x \geq -\dfrac{4}{3}$ such as 0 into the original equation.

$$14x + 4 \leq 26x + 20$$
$$14(0) + 4 \leq 26(0) + 20$$
$$4 \leq 20 \text{ True}$$

SOLVING A LINEAR INEQUALITY

1. Solve the inequality: $-2x + 3 < 6$

    a. $x < -\dfrac{3}{2}$ b. $x > -\dfrac{9}{2}$ c. $x < \dfrac{3}{2}$ d. $x > -\dfrac{3}{2}$

2. Solve the inequality: $2(9x + 7) < 3$

    a. $x > -\dfrac{11}{18}$ b. $x < \dfrac{1}{18}$ c. $x > \dfrac{1}{18}$ d. $x < -\dfrac{11}{18}$

3. Solve the inequality: $19x + 4 < 51x + 1$

    a. $x > -\dfrac{3}{32}$ b. $x < -\dfrac{3}{32}$ c. $x > \dfrac{3}{32}$ d. $x < \dfrac{3}{32}$

4. Solve the inequality: $13x + 5 < 36x + 9$

    a. $x > \dfrac{4}{23}$ b. $x > -\dfrac{4}{23}$ c. $x < \dfrac{4}{23}$ d. $x < -\dfrac{4}{23}$

SOLUTIONS:

1. d   2. d   3. c   4. b

IDENTIFY INTERCEPTS OF A LINEAR (AX + BY = C)

29. Find the y –intercept for: $-5x + 8y = -8$

A. $\left(\dfrac{8}{5}, 0\right)$ B. $\left(0, \dfrac{8}{5}\right)$ C. $(0, -1)$ D. $\left(\dfrac{8}{5}, -1\right)$

Solution:

Find the y –intercept for: $-5x + 8y = -8$.

To find the x–intercept of a linear equation make $y = 0$ and solve for x. To find the y–intercept of a linear equation make $x = 0$ and solve for y.

The first step to find the y–intercept is to substitute 0 for x and solve for y.

x	y
0	?

$-5x + 8y = -8$
$-5(0) + 8y = -8$
$8y = -8$
$\div 8 \quad \div 8$
$y = -1$

x	y
0	-1

The second step is to write the solution as a coordinate (x, y). The value for x is 0 and the value for y is –1.

The solution is: $(0, -1)$

IDENTIFY INTERCEPTS OF A LINEAR (ax + by = c)

1. Find the x-intercept for: $-5x + 4y = -3$

   a. $\left(0, \dfrac{3}{5}\right)$ b. $\left(0, -\dfrac{3}{4}\right)$ c. $\left(\dfrac{3}{5}, 0\right)$ d. $\left(\dfrac{3}{5}, -\dfrac{3}{4}\right)$

2. Find the y-intercept for: $-2x - 7y = 3$

   a. $\left(0, -\dfrac{3}{7}\right)$ b. $\left(-\dfrac{3}{2}, 0\right)$ c. $\left(0, -\dfrac{3}{2}\right)$ d. $\left(-\dfrac{3}{2}, -\dfrac{3}{7}\right)$

3. Find the y-intercept for: $9x + 5y = 3$

   a. $\left(\dfrac{1}{3}, \dfrac{3}{5}\right)$ b. $\left(0, \dfrac{1}{3}\right)$ c. $\left(0, \dfrac{3}{5}\right)$ d. $\left(\dfrac{1}{3}, 0\right)$

4. Find the x-intercept and y-intercept for: $-5x + 4y = -8$

   a. x-intercept: $\left(0, \dfrac{8}{5}\right)$ y-intercept: $(-2, 0)$ b. x-intercept: $\left(\dfrac{8}{5}, 0\right)$ y-intercept: $(0, -2)$

   c. x-intercept: $(0, -2)$ y-intercept: $\left(\dfrac{8}{5}, 0\right)$ d. x-intercept: $(-2, 0)$ y-intercept: $\left(0, \dfrac{8}{5}\right)$

SOLUTIONS:

1. c 2. a 3. c 4. b

PRACTICE PROBLEMS

MATCH LINEAR EQUATION TO GRAPH

30. Find the graph that best matches the linear equation: $y = \frac{1}{2}x - 4$.

A
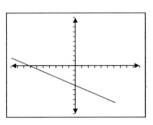

B
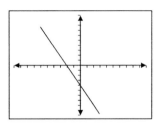

C
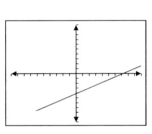

D
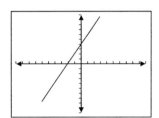

One method is to create a table of values. Choose a value for x or y, then solve for the other variable. Three points are sufficient for a linear equation.

x	y
0	?
2	?
4	?

$y = \frac{1}{2}(0) - 4$
$y = -4$

$y = \frac{1}{2}(2) - 4$
$y = 1 - 4$
$y = -3$

$y = \frac{1}{2}(4) - 4$
$y = 2 - 4$
$y = -2$

x	y	(x, y)
0	-4	(0, -4)
2	-3	(2, -3)
4	-2	(4, -2)

Label the points and draw a line through the points.

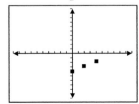
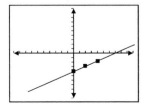

Another method is to use the slope intercept form (y = mx + b) to graph the equation.

360 PRACTICE PROBLEMS

The first step is to make sure that the linear equation is written in an explicit form:

$$y = \frac{1}{2}x - 4$$

The next step is to determine the slope and y-intercept of the linear equation.

In the form: y = mx + b, the m is the slope and the b is the y intercept.

Slope (m) = $\frac{1}{2}$ and y intercept (b) = –4

To graph, first place a point on the y intercept at –4. The y intercept is the point where the graph crosses the y axis. In this case the y intercept is –4.

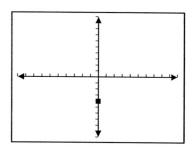

The next step is to use the slope to find the next point. The slope can be expressed as $\frac{rise}{run}$. The numerator determines if you move up (positive value) or down (negative value). The denominator determines if you move left (negative value) or right (positive value).

In this case, Slope (m) = $\frac{1}{2}$ means move up 1 unit and to the right 2 units.

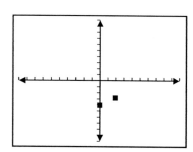

Then draw a straight line connecting the points.

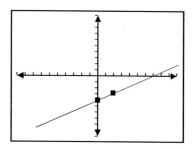

This is the solution.

PRACTICE PROBLEMS

MATCH LINEAR EQUATION TO GRAPH

1. Find the graph that best matches the given linear equation: $y = 4x + 3$

    a.
    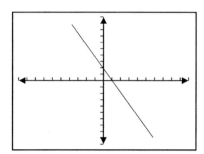

    b.
    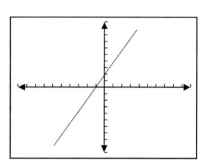

    c.
    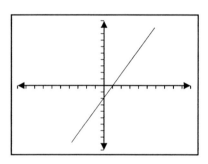

    d.
    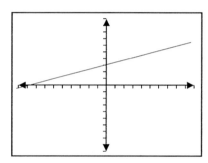

2. Find the graph that best matches the given linear equation: $2x + y = 1$

    a.
    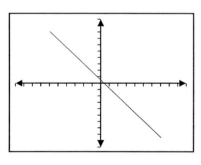

    b.
    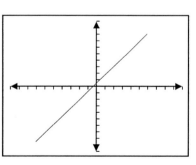

    c
    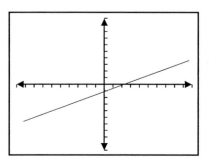

    d.
    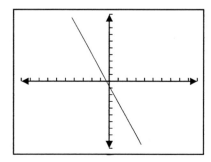

PRACTICE PROBLEMS

3. Find the graph that best matches the given linear equation: $y = -\dfrac{2}{3}x - 2$

a.
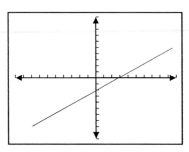

b.
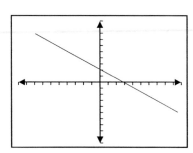

c.
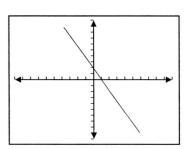

d.
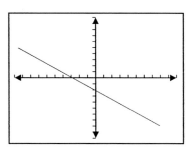

4. Find the graph that best matches the given linear equation: $2x + y = 1$

a.
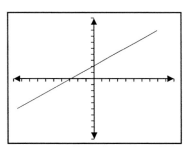

b.
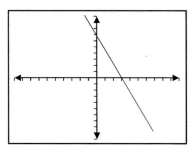

c
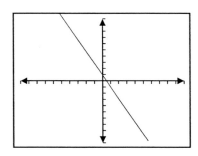

d.
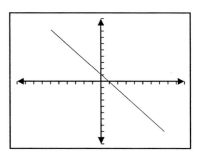

SOLUTIONS:

1. b   2. a   3. d   4. c

PRACTICE TEST

1. Simplify: $8 - 14 \div 7 + 9$
   a. 19
   b. 15
   c. $\dfrac{57}{8}$
   d. $-\dfrac{3}{8}$

2. Simplify: $30 - 4^2 \div 2 - 9 \cdot 4$
   a. -14
   b. $\dfrac{77}{2}$
   c. $\dfrac{326}{11}$
   d. 286

3. Simplify: $|-22|+7|-|-15|$
   a. -4
   b. 14
   c. 30
   d. 44

4. Simplify: $-6[6x + 30 + x]$
   a. $-42x - 180$
   b. $30x + 180$
   c. $-42x + 180$
   d. $42x + 180$

5. Evaluate the given expression when $x = -2$, $y = 3$.
   $6x^2 - 3y^2$
   a. -51
   b. -3
   c. 6
   d. 63

6. Solve for y: $-9 - 3y - 6 = 7y + 14$
   a. $y = -\dfrac{29}{17}$
   b. $y = \dfrac{25}{17}$
   c. $y = \dfrac{5}{2}$
   d. $y = -\dfrac{29}{10}$

7. Solve for x: $\dfrac{2}{3}x - \dfrac{1}{2}x = -\dfrac{1}{6}x - 2$
   a. $x = -12$
   b. $x = -1$
   c. $x = -6$
   d. $x = -2$

8. Solve for y: $t = -5x + 3y$

   a. $y = \dfrac{1}{3}t - 5x$
   b. $y = \dfrac{1}{3}t - 5x$
   c. $y = \dfrac{1}{3}t + \dfrac{5}{3}x$
   d. $y = \dfrac{1}{3}t - \dfrac{5}{3}x$

9. The product of two consecutive even integers is 48. Choose the equation that could be used to find this number, x.

   a. $x + (x+2) = 48$
   b. $x(x+2) = 48$
   c. $x(x+1) = 48$
   d. $(x+1)(x+2) = 48$

10.1. If a new TV set costs $326 after a 20% discount, what was the original price?

   a. $407.50
   b. $1630.00
   c. $260.80
   d. $65.20

10.2. The perimeter of a triangle is 30 inches. The length of the middle side is 2 inches more than the length of the smallest side and the largest side is 4 inches more than twice the length of the smallest side. Find the length of the smallest side.

   a. 6 inches
   b. 8 inches
   c. 4 inches
   d. 2 inches

10.3. Two tuna boats start from the same port at the same time, but they head to opposite directions. The faster boat travels 10 knots per hour faster than the slower boat. At the end of 8 hours, they were 272 nautical miles apart. How many nautical miles had each boat traveled by the end of the 8-hour period?

   a. 12 and 22 nautical miles
   b. 96 and 176 nautical miles
   c. 131 and 141 nautical miles
   d. 17 and 27 nautical miles

11. Identify the proportion listed below that solves the problem: If 6 apples cost $1.38, how much will 16 apples cost?

   a. $\dfrac{6}{16} = \dfrac{x}{1.38}$
   b. $\dfrac{6}{1.38} = \dfrac{x}{16}$
   c. $\dfrac{x}{16} = \dfrac{6}{1.38}$
   d. $\dfrac{6}{16} = \dfrac{1.38}{x}$

12. Simplify: $\dfrac{(2x^5 y^2)^3}{y^5}$

   a. $8x^{20}y$
   b. $8x^{15}y$
   c. $2x^{15}y^2$
   d. $2x^{20}y^6$

13. Simplify: $\dfrac{x^4 y^{-2}}{x^{-4} y^5}$

    a. $\dfrac{1}{x^8} y^7$
    b. $\dfrac{1}{x^8 y^7}$
    c. $\dfrac{x^8}{y^7}$
    d. $x^8 y^7$

14. Simplify: $\left(x^5 y^0\right)^{-4}$

    a. $\dfrac{1}{x^{20}}$
    b. $x^9 y^4$
    c. $\dfrac{x}{y^4}$
    d. x^{20}

15.1. Convert to scientific notation: 0.00000427

    a. 4.27×10^6
    b. 0.427×10^{-5}
    c. 0.427×10^{-5}
    d. 4.27×10^{-5}

15.2 Convert to standard form: 3.54×10^8

    a. 0.00000000354
    b. 354,000,000
    c. 3,540,000,000
    d. 0.000000354

16. Simplify: $(9x^2 + 6x - 7) - (2x^2 - 7x + 5)$

    a. $7x^2 + 13x - 2$
    b. $7x^2 - x - 2$
    c. $7x^2 - x - 12$
    d. $7x^2 + 13x - 12$

17. Simplify: $3x(-4x - 5)$

    a. $-12x^2 + 15x$
    b. $-60x^2$
    c. $12x^2 - 15x$
    d. $-12x^2 - 15x$

18. Simplify: $(-7x + 6)(7x + 2)$

    a. $-49x^2 - 28x - 12$
    b. $-49x^2 + 56x + 12$
    c. $-49x^2 + 28x + 12$
    d. $-49x^2 + 56x - 12$

19. Factor completely: $60w^8v^{12} - 40w^{10}v^3$
    a. $20w^8v^3\left(3v^9 - 2w^2\right)$
    b. $20w^8v^3\left(3w^9 - 2v^2\right)$
    c. $20w^2v^3\left(3w^4v^4 - 2w^5v\right)$
    d. $10w^8v^3\left(6v^9 - 4w^2\right)$

20. Factor completely: $9s^2 - 25t^2$
    a. $(3s + 5t)(3s + 5t)$
    b. $(3s - 5t)(3s - 5t)$
    c. $(3s - 5t)(3s + 5t)$
    d. $(3t - 5s)(3t + 5s)$

21. Factor completely: $-4t - 4s - 4tz - 4sz$
    a. $-4(t + s)(1 + z)$
    b. $4(t - s)(1 - z)$
    c. $(t - s)(-4 - 4z)$
    d. $-4(t + s)(1 + z)$

22. Identify a factor of the trinomial: $3x^2 - 11x + 10$
    a. $(x - 2)$
    b. $(x + 2)$
    c. $(3x + 5)$
    d. $(3x - 2)$

23. Simplify: $\dfrac{2x^2 + 5x - 3}{9 - x^2}$
    a. $\dfrac{2x - 1}{x - 3}$
    b. $\dfrac{1 - 2x}{3 - x}$
    c. $\dfrac{2x - 1}{3 - x}$
    d. $\dfrac{2x + 1}{x - 3}$

24. Solve: $x^2 - 4x - 5 = 0$
    a. $x = 5, x = 1$
    b. $x = \dfrac{1}{5}, x = -1$
    c. $x = -5, x = 1$
    d. $x = 5, x = -1$

25. Solve: $4x^2 - 4x - 35 = 0$
    a. $x = -\dfrac{7}{2}, x = -\dfrac{5}{2}$
    b. $x = -\dfrac{7}{2}, x = \dfrac{5}{2}$
    c. $x = \dfrac{2}{7}, x = \dfrac{5}{2}$
    d. $x = \dfrac{7}{2}, x = -\dfrac{5}{2}$

26. Assuming the variable represents a non-negative number, simplify completely: $\sqrt[3]{25x^7z^8}$
    a. $75z^4x^3\sqrt{x^7}$
    b. $15z^4\sqrt{x^7}$
    c. $75z^4x^3\sqrt{x}$
    d. $15z^4x^3\sqrt{x}$

27.1. Simplify completely: $\sqrt{2}\left(\sqrt{7}-2\sqrt{8}\right)$

a. $\sqrt{14}+8$
b. $\sqrt{14}-\sqrt{32}$
c. $\sqrt{4}-4$
d. $-8+\sqrt{14}$

27.2. Simplify completely: $\left(2\sqrt{3}-7\right)\left(\sqrt{3}+7\right)$

a. $2\sqrt{9}-49$
b. $-43-7\sqrt{3}$
c. $-43+7\sqrt{3}$
d. $2\sqrt{3}-14$

28. Solve the inequality: $6(9y+7)<5$

a. $y<-\dfrac{37}{54}$
b. $y>-\dfrac{1}{54}$
c. $y>-\dfrac{37}{54}$
d. $y<-\dfrac{1}{54}$

29.1. Find the x-intercept for: $-9x - 5y = 4$

a. $\left(-\dfrac{4}{9},-\dfrac{4}{5}\right)$
b. $\left(0,-\dfrac{4}{9}\right)$
c. $\left(-\dfrac{4}{5},0\right)$
d. $\left(-\dfrac{4}{9},0\right)$

29.2. Find the y-intercept for: $9x + 8y = 9$

a. $\left(0,\dfrac{9}{8}\right)$
b. $(1, 0)$
c. $(0, 1)$
d. $\left(1,\dfrac{9}{8}\right)$

30. Find the graph that best matches the given linear equation: $y = x - 7$

368 PRACTICE PROBLEMS

a. b.

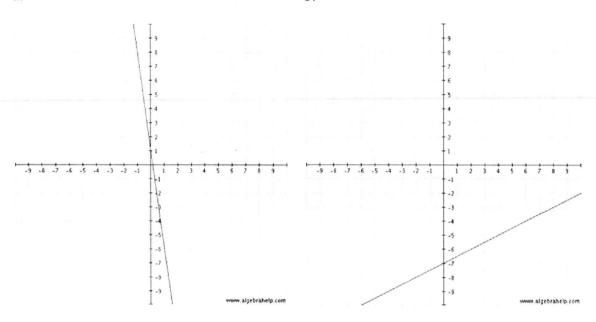

c. d.

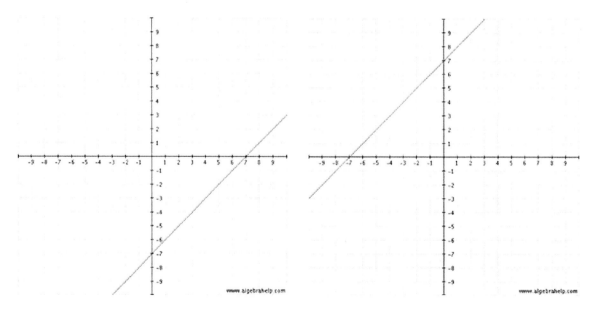

SAMPLE TEST – SOLUTIONS

1. B
2. A
3. B
4. A
5. B
6. D
7. C
8. C
9. B
10.1. A
10.2. A
10.3 B
11. D
12. C
13. C
14. A
15.1. C
15.2. B
16. D
17. D
18. C
19. A
20. C
21. A
22. A
23. C
24. D
25. D
26. D
27.1. D
27.2. C
28. A
29.1. D
29.2. A
30. C

PRACTICE PROBLEMS

SAMPLE EXAM

1. Simplify: 6 - 18 ÷ 9 - 4

   A. 19   B. 34   C. 7   D. 0

   1. _____

2. Simplify: $23 - (8)^2 \div (14 - 6) \cdot 6$

   A. −25   B. 90   C. $-\dfrac{123}{4}$ D. $\dfrac{65}{3}$

   2. _____

3. Simplify: $|8 + (-14)| + 9$

   A. 16   B. 31   C. 3   D. 15

   3. _____

4. Simplify: $-6[6(x+5)+x]$

   A. 42x - 180   B. 30x + 180   C. -42x - 180   D. 42x + 180

   4. _____

5. Evaluate the given expression when w = − 4: $-6w^2 + 5w + 3$

   A. − 79   B. − 119   C. − 113   D. − 73

   5. _____

6. Solve for r: $-2(-9r - 6) = 6(r + 5)$

   A. $r = \dfrac{7}{2}$ B. $r = -\dfrac{3}{2}$ C. $r = \dfrac{7}{4}$ D. $r = \dfrac{3}{2}$

   6. _____

7. Solve for y: $\frac{6}{5}y - 6 = -6$　　　　　7. _____

　A.　y = −10　　B.　y = 0　　C.　y = − 6　　D.　y = − 4

8. Solve for t: $x = -8z + 7t$　　　　　8. _____

　A.　$t = \frac{1}{7}x - 8z$　　B.　$t = \frac{1}{7}x + \frac{8}{7}z$　　C.　$t = \frac{1}{7}x + 8z$　　D.　$t = \frac{1}{7}x - \frac{8}{7}z$

9. The sum of a number and 16 is 4 more than twice the number.
   Find the equation that could be used to find this number, x.　　9. _____

　A.　$x + 16 = x^2 + 4$　　B.　$16x = 2x + 4$　　C.　$x + 16 = 2(x + 4)$　　D.　$x + 16 = 2x + 4$

10. The length of a rectangle is 2 feet more than the width. The perimeter of the rectangle is 72 feet. Find the length.　　10. _____

　A.　19 feet　　B.　17 feet　　C.　37 feet　　D.　35 feet

11. Identify the proportion listed below that solves this problem:
    A car can travel 500 miles on 5 gallons of gasoline.
    How far can the car travel on 32 gallons of gasoline?　　11. _____

　A.　$\frac{5}{500} = \frac{x}{32}$　　B.　$\frac{5}{500} = \frac{32}{x}$　　C.　$\frac{5}{32} = \frac{x}{500}$　　D.　$\frac{500}{5} = \frac{32}{x}$

PRACTICE PROBLEMS

12. Simplify: $(a^2b^4)^3(a^3b^4)$

    A. $a^{10}b^{12}$ B. a^9b^{16} C. a^6b^{16} D. a^6b^{12}

    12. _____

13. Simplify: $\dfrac{a^{-5}b^{-3}}{a^5b^5}$

    A. $\dfrac{a^{10}}{b^8}$ B. $\dfrac{1}{a^{10}b^8}$ C. $a^{10}b^8$ D. $\dfrac{b^8}{a^{10}}$

    13. _____

14. Simplify: $(a^2b^{-4}c^0)^{-3}$

    A. $a^{-1}b^{-7}c^{-3}$ B. $\dfrac{b^{12}}{a^6c^3}$ C. $\dfrac{b^{12}}{a^6}$ D. $\dfrac{a^6}{b^{12}}$

    14. _____

15. Convert to scientific notation: .000000195

    A. 0.195×10^{-6} B. 1.95×10^{-7} C. 1.95×10^7 D. 1.95×10^{-6}

    15. _____

16. Simplify: $(7x^2 - 8x + 4) - (6x^2 - 8x - 5)$

    A. $x^2 - 16x - 1$ B. $x^2 - 1$ C. $x^2 + 1$ D. $x^2 + 9$

    16. _____

17. Simplify: $-7x(-4x+7)$ 17. _____

   A. $28x^2 - 49x$ B. $28x^2 + 49x$ C. $-28x^2 - 49x$ D. $-21x^2$

18. Simplify: $(4x-7)(6x-7)$ 18. _____

   A. $24x^2 + 14x + 49$ B. $24x^2 - 70x + 49$ C. $24x^2 - 14x - 49$ D. $24x^2 - 70x - 49$

19. Factor completely: $30y^{50}s^8 + 14y^{55}s^6 + 24y^{15}s^6$ 19. _____

   A. $2y^5s^2\left(15y^{10}s^4 + 7y^{11}s^3 + 12y^3s^3\right)$
   B. $2ys\left(15y^{49}s^7 + 7y^{54}s^5 + 12y^{14}s^5\right)$
   C. $2y^{14}s^5\left(15y^{36}s^3 + 7y^{41}s^1 + 12y^1s^1\right)$
   D. $2y^{15}s^6\left(15y^{35}s^2 + 7y^{40} + 12\right)$

20. Factor completely: $25z^2 - 64x^2$ 20. _____

   A. $(5z + 16x)(5z - 4x)$
   B. $(5z - 8x)^2$
   C. $(5z + 8x)(5z - 8x)$
   D. $(5z + 4x)(5z - 16x)$

21. Factor completely: $yr + yk + br + bk$ 21. _____

   A. $(y-b)(r-k)$ B. $(y+b)(r+k)$ C. $(y+b)(r-k)$ D. $(y-b)(r+k)$

PRACTICE PROBLEMS

22. Identify a factor of the following trinomial: $6t^2 - 25t + 4$

    A. $(3t+2)$ B. $(3t-2)$ C. $(t-4)$ D. $(t-1)$

    22. _____

23. Simplify: $\dfrac{2x^2 - 5x + 3}{x^2 - 1}$

    A. $\dfrac{2x-3}{x+1}$ B. $\dfrac{2x+3}{x+1}$ C. $\dfrac{x-2}{x+1}$ D. $\dfrac{2x-3}{2x-3}$

    23. _____

24. Solve: $x^2 + 6x + 5 = 0$

    A. $x = \dfrac{1}{5},\ x = -1$ B. $x = -5,\ x = 1$ C. $x = -5,\ x = -1$ D. $x = 5,\ x = 1$

    24. _____

25. Solve: $35x^2 + 11x - 6 = 0$

    A. $x = -\dfrac{3}{5},\ x = -\dfrac{2}{7}$ B. $x = \dfrac{5}{3},\ x = \dfrac{2}{7}$ C. $x = \dfrac{3}{5},\ x = -\dfrac{2}{7}$ D. $x = -\dfrac{3}{5},\ x = \dfrac{2}{7}$

    25. _____

26. Simplify completely: $\sqrt{18x^7 w^4}$

    A. $9w^2 x^3 \sqrt{2x^7}$ B. $9w^2 x^3 \sqrt{2x}$ C. $3w^2 \sqrt{2x^7}$ D. $3w^2 x^3 \sqrt{2x}$

    26. _____

27. Simplify: $3\sqrt{2}\left(\sqrt{10} - 5\sqrt{2}\right)$ 27. _____

    A. $3\sqrt{20} - 60$ B. $-24\sqrt{5}$ C. $12\sqrt{5} - 15\sqrt{4}$ D. $6\sqrt{5} - 30$

28. Solve the inequality: $14x + 4 \leq 26x + 20$ 28. _____

    A. $x \geq \dfrac{4}{3}$ B. $x \leq \dfrac{4}{3}$ C. $x \geq -\dfrac{4}{3}$ D. $x \leq -\dfrac{4}{3}$

29. Find the y–intercept for: $-5x + 8y = -8$ 29. _____

    A. $\left(\dfrac{8}{5}, 0\right)$ B. $\left(0, \dfrac{8}{5}\right)$ C. $(0, -1)$ D. $\left(\dfrac{8}{5}, -1\right)$

30. Find the graph that best matches the linear equation: $y = \dfrac{1}{2}x - 4$ 30. _____

A.

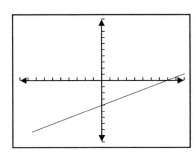

B.

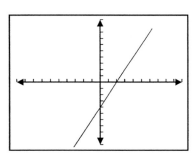

C.

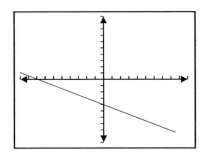

D.

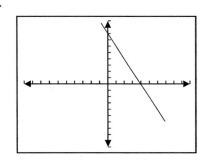

ANSWERS SAMPLE EXAM

1. D
2. A
3. D
4. C
5. C
6. D
7. B
8. B
9. D
10. A
11. B
12. B
13. B
14. C
15. B
16. D
17. A
18. B
19. D
20. C
21. B
22. C
23. A
24. C
25. D
26. D
27. D
28. C
29. C
30. A